MathWise Percents

With Answer Key

Peter Wise

CONTRIBUTORS

David Wise

Katherine Wise

Cover Design by Kris Budi

Dedicated to my wife, Allison, and to my two children, David and Katherine.

Special thanks to Shi Hayes for her helpful suggestions and for proofreading the introduction, and to Aileen Finnegan for her leadership and support.

MathWise Percents with Answer Key

MathWise Curriculum Press

First printing 2014

MathWise Percents

TABLE OF CONTENTS

FOR WHICH GRADE LEVEL(S) ARE THESE BOOKS INTENDED?

This series is based on skill sets, not grade or age. These workbooks are intentionally created to be suitable for a wide range of grades. They are focused on math topics, irrespective of grade level or age. If students in any grade need extra support in a given topic area, these books are designed to enrich whatever curriculum they may be using. If a student in middle or high school student is rusty on a skill set normally taught in earlier grades, this series will help. There should be no stigma attached to reviewing important content in math, language arts, or even a foreign language. On the other hand, if a student in 3rd grade has a parent or teacher who is willing to walk the student through the explanations and exercises in this book, he or she will also profit from the study. Front-loading key math concepts will make future math classes that much easier.

These books were part of the instruction in several different grade levels, and even in multi-grade math clubs. No one gets distracted by the grade level of the material. The concepts are the target.

MY EMPHASES IN TEACHING MATH

Too many students learn math as if they were learning a dead language. To them math consists of memorizing a bunch of rules and formulas. This is the wrong approach to learning math. To be good at math, it is important to know **how and why math works the way it does.** Students need to be trained to think mathematically from preschool through college, in every grade level. A formulaic understanding of math is both harder to learn and easier to forget.

Tips and tricks help as memory aids and have a legitimate role in acquiring and retaining information. However it is even more important that students understand the reasoning behind rules and formulas. **The MathWise series incorporates both tips/tricks as well as reasoning behind math formulas and procedures.**

THE FORMAT OF THIS BOOK

The explanations, graphics, and format of this series is designed to be kid-friendly, upbeat, and as appealing as possible. I have incorporated tips, tricks, and other pedagogical secrets into this book. Students tell me that they like the format and self-contained explanations. Every year students make breakthroughs in their understanding of math through these pages.

HOW TO USE THIS BOOK

While students can use this series profitably when working alone, my experience indicates that they will make greater progress if guided by a parent, tutor, or helper. This is particularly true for younger students. This person need not be a math teacher at all—just a reader.

If a student or parent is unclear about a solution or procedure, he or she is encouraged to check with the answer key.

Web Site
For questions, comments, or suggestions, please visit www.mathwisebooks.com.

SPECIAL FEATURES OF THIS BOOK

This series of books is designed to be unique and to catch kids' attention in special ways:

Tips and Tricks
Over the years, I have assembled a wide assortment of memory aids—my tips and tricks. Students have found these to be helpful and memorable, but they have also found that these pointers add interest and excitement to their math studies.

Speech Bubbles with Teacher Insights
Speech bubbles are used to provide guidance, point out insights, or give helpful hints as students are solving math problems. Students learn best by doing, and the instruction given in the speech bubbles is designed to (1) sharpen students' powers of observation, (2) increase number sense, and (3) instruct in digestible chunks.

Higher-Order Mathematical Thinking Skills
Rather than depending solely on rote memorization, this series endeavors to explain to students how and why the math does what it does all along the way. This increases retention of concepts and fosters a deeper understanding of mathematics

Simplicity of Instruction
Concepts are explained clearly and simply. Every page (excluding review pages or concept quizzes) has a specific focus. Most pages have generous amounts of white space to keep students focused. Movement is from the simple to the increasingly complex.

Step-By-Step Procedures
Students learn best when given explicit, step-by-step instruction. When several steps are involved, they are numbered. This makes learning much more logical and memorable.

Depth and Complexity
Throughout the book there are challenge problems to stretch students' thinking. At your discretion, you can guide students through the more challenging problems (recommended) or, alternatively, you can skip these harder problems.

Informal Terms
This book often employs informal language like "top number" or "bottom number" to keep things simple and focused. Standard mathematical terminology, such as numerator and denominator, is used after the concepts are presented.

Logical-Sequential Instruction
Math problems are presented in a logical sequence so that previous problems contribute to students' abilities to solve future problems. The order in which you present math problems is critical to promoting number sense.

TEACHING PERCENTS

Percents and Number Sense

Percents help students strengthen their understanding of number sense. They not only involve basic operations, they also utilize decimal skills, fraction skills, and manipulation of numbers. For instance, when computing 70% of a number, a student may first find 10% of the number, and then multiply by 7. When calculating a 15% tip, he or she may first find 10%, then cut this amount in half and add it to the 10% (i.e., 15% = 10% + 5%). 60% may also be calculated by finding the sum of 50% + 10%. I have seen firsthand, how easily students of all ability levels have come up with clever alternative ways of computing percents.

Percents and the Real World

Percents are seen everywhere in stores and ads—these can provide helpful opportunities for reinforcing percent skills. There are other fun, practical, and motivating ways to practice percents, such as: test scores ("Seven out of eight correct is what percent?"), rate of change, ("Tom got 3 baskets in his first basketball game, then he got 5 on his second game; what percent was his increase?"), and sports statistics.

Importance of Fraction-Decimal-Percent Conversions

I often tell my students that they can think of numbers as wearing three different costumes: fraction costumes, decimal costumes, and percent costumes. The exact same "actor" (number) can slip into different costumes, but he remains the same person. Students need to be able to put different outfits on numbers; that is, to convert fractions to decimals, decimals to percents, and so on. Proficiency in these conversions is an integral part of all math.

What are Percents Anyway?

Percents are amounts out of 100

A. Percent means "out of a hundred"

per = out of

cent = hundred

100 cents in a dollar — century = 100 years

B. A PERCENT SIGN IS A REARRANGED 100!

COMPARE!

100 %

C. $\frac{7}{10} = \frac{70}{100}$ When a number is over 100 or out of 100, it is a percent

A NUMBER OVER 100 IS THE SAME THING AS A PERCENT!

7 out of 10 — 70 out of 100

= 70 %

D.

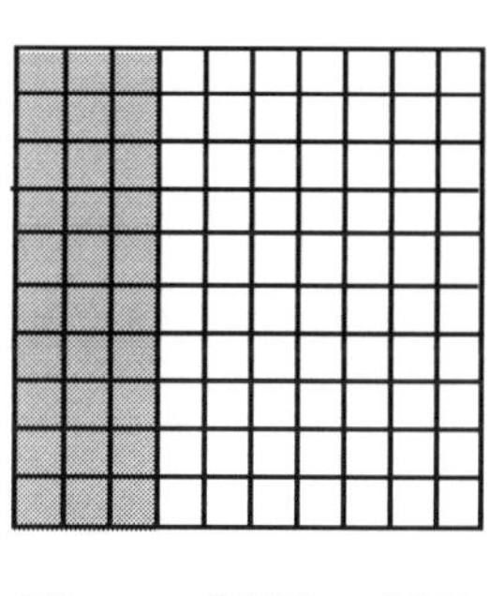

THIS SQUARE IS DIVIDED INTO 100 BOXES

A HUNDREDS SQUARE IS A GOOD VISUAL DEMONSTRATION OF A PERCENT!

30 out of 100 is 30%

E. PERCENTS CAN BE WRITTEN AS DECIMALS!

35% is the same as .35

TWO DECIMAL SLIDES FOR THE TWO ZEROS IN THE PERCENT SIGN!

F. WHOLE NUMBERS CAN BE WRITTEN AS PERCENTS!

100% = 1

400% = 4

Visualizing Percents

Example

A. Amount shaded: 70% Amount not shaded: 30%

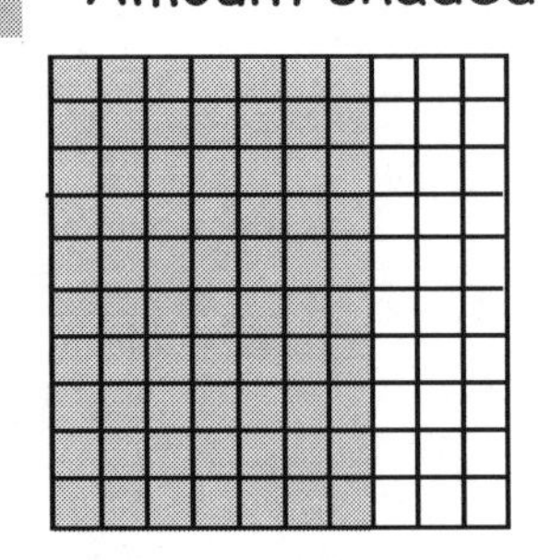

These two amounts will always add up to 100%

70 out of 100 is 70%

Tell or shade the correct amounts

1.

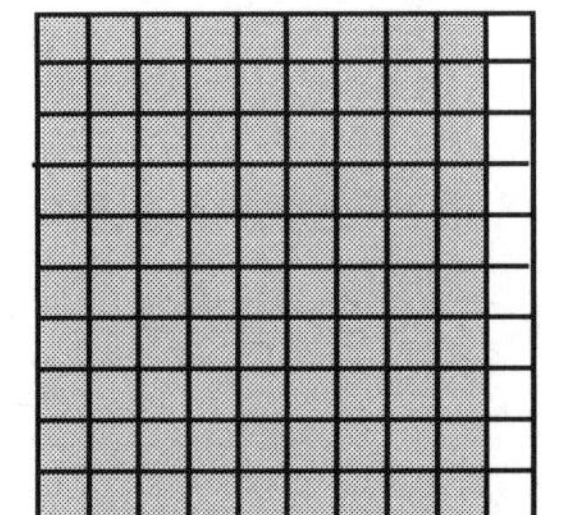

Shaded:

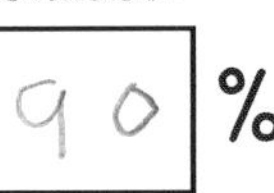
%

Not shaded:

10 %

2.

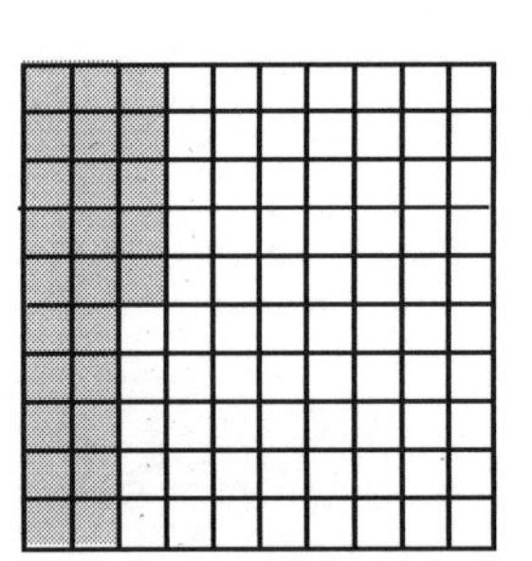

Shaded:

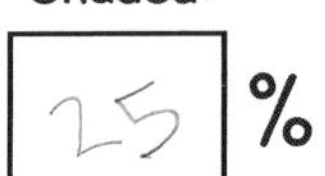
%

Not shaded:

75 %

3.

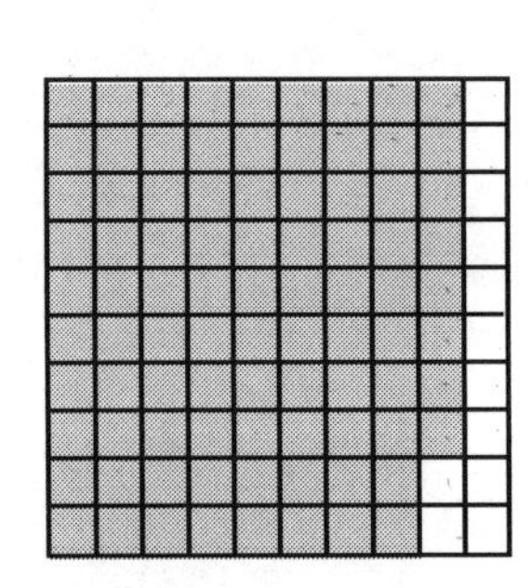

Shaded:

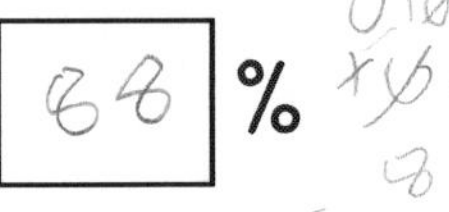
%

Not shaded:

12 %

4.

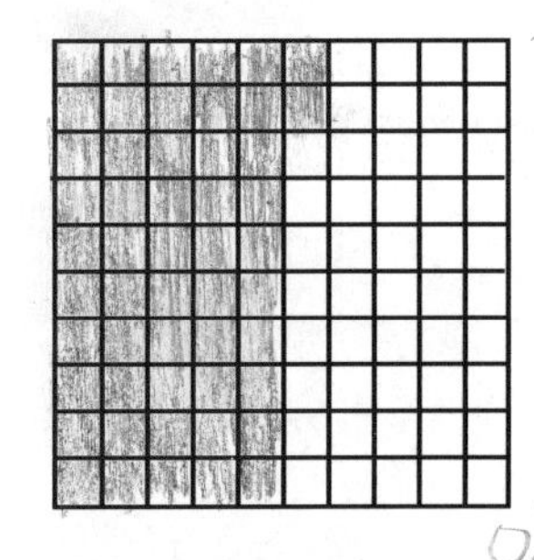

Shaded:

52 %

Not shaded:

%

5.

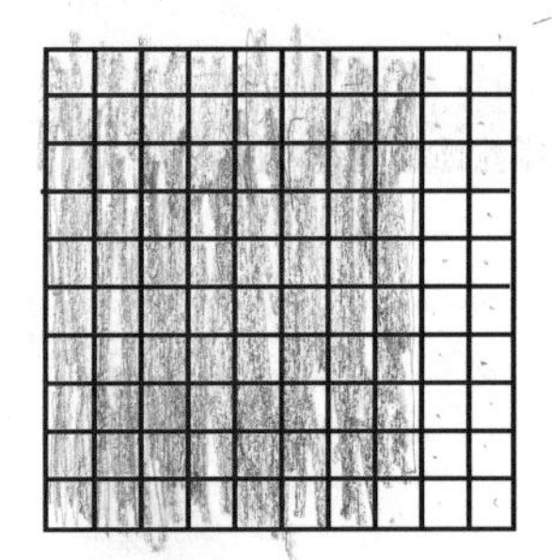

Shaded:

79 %

Not shaded:

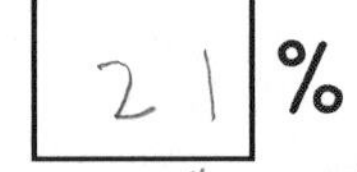
%

6.

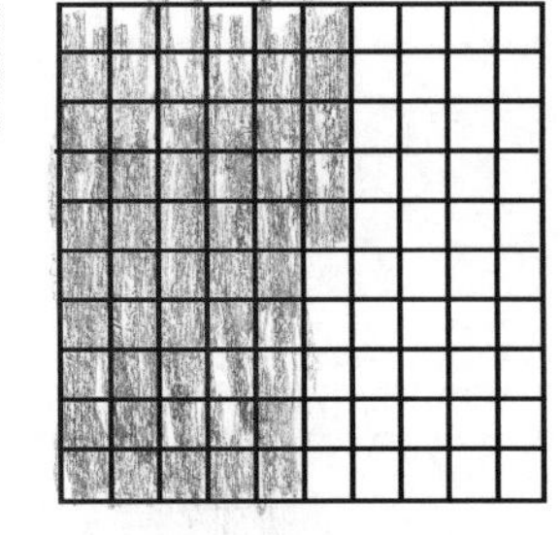

Shaded:

%

Not shaded:

35 %

Determining Basic Percents

Divide into 100 to figure out the following percents

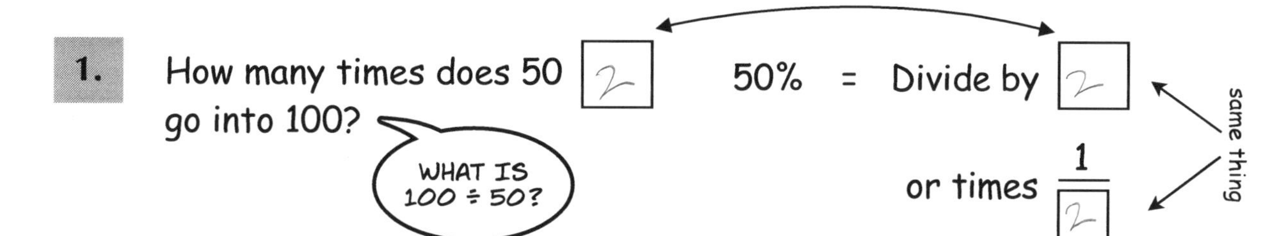

1. How many times does 50 go into 100? [2] 50% = Divide by [2] or times $\frac{1}{[2]}$

2. How many times does 25 go into 100? [4] 25% = Divide by [4] or times $\frac{1}{[4]}$

3. How many times does 20 go into 100? [5] 20% = Divide by [5] or times $\frac{1}{[5]}$

4. How many times does 10 go into 100? [10] 10% = Divide by [10] or times $\frac{1}{[10]}$

or one decimal slide to the LEFT:

10% of 40 = 4.0 or 4

5. How many times does 5 go into 100? [20] 5% = Divide by [20] or times $\frac{1}{[20]}$

6. How many times does 1 go into 100? [100] 1% = Divide by [100] or times $\frac{1}{[100]}$

or two decimal slides to the LEFT

1% of 70 = .70 or .7

Basic Percent Values

Percents to Memorize

A. 100% = Divide by 1 or times 1

B. 50% = Divide by 2 or times $\frac{1}{2}$

C. 25% = Divide by 4 or times $\frac{1}{4}$

D. 20% = Divide by 5 or times $\frac{1}{5}$

E. 10% = Divide by 10 or times $\frac{1}{10}$

F. 1% = Divide by 100 or times $\frac{1}{100}$

G. 5% = Divide by 20 or times $\frac{1}{20}$

H. 15% = 10% + 5% (half of 10%)

I. $33.\overline{3}\%$ = Divide by 3 or times $\frac{1}{3}$

$33\frac{1}{3}\%$

J. $66.\overline{6}\%$ = Divide by 3 and times 2

$66\frac{2}{3}\%$ or times $\frac{2}{3}$ (2 ← ×, 3 ← ÷)

K. 75% = Divide by 4 and times 3

or times $\frac{3}{4}$ (3 ← ×, 4 ← ÷)

Calculate the percentages of the following numbers

1. 100% of 45 = ☐

2. 50% of 60 = ☐

3. 25% of 12 = ☐

4. 20% of 35 = ☐

5. $33.\overline{3}\%$ of 18 = ☐

6. $66.\overline{6}\%$ of 24 = ☐

7. 10% of 90 = ☐

8. 1% of 80 = ☐

9. 5% of 80 = ☐

10. 25% of 32 = ☐

11. 20% of 40 = ☐

12. 10% of 40 = ☐

Percent Practice

Calculate the percentages of the following numbers

1. 50% of 40 = ☐

25% IS HALF OF 50%!

2. 25% of 40 = ☐

75% IS JUST 50% + 25%.

...SO JUST ADD THE PREVIOUS TWO!

3. 75% of 40 = ☐

4. 50% of 36 = ☐

5. 25% of 36 = ☐

6. 75% of 36 = ☐

7. 20% of 55 = ☐

8. 10% of 60 = ☐

+ +

9. 5% of 60 = ☐

= = ↓

10. 15% of 60 = ☐

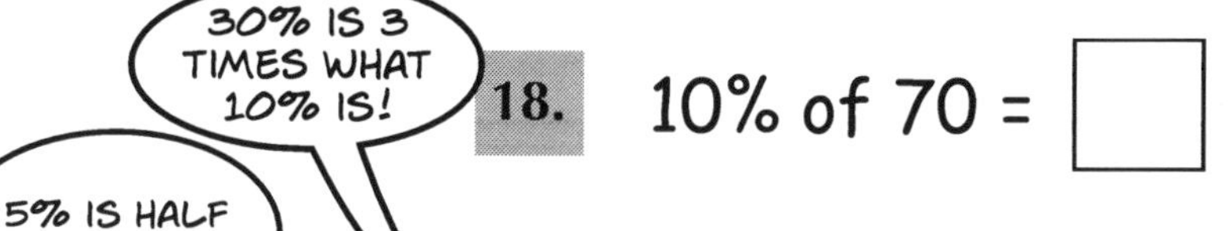

30% IS 3 TIMES WHAT 10% IS!
5% IS HALF OF 10%!

15% IS 10% + 5%!

11. 10% of 20 = ☐

12. 50% of 26 = ☐

13. 25% of 32 = ☐

14. 20% of 15 = ☐

15. 10% of 90 = ☐

16. 25% of 24 = ☐

17. 20% of 60 = ☐

18. 10% of 70 = ☐

19. 30% of 70 = ☐

PERCENT = DIVIDE BY 100!

THE TWO ZEROS ON THE NUMBERS MEAN "TIMES 100"!

20. 80% of 70 = ☐

80% IS 8 TIMES WHAT 10% IS!

NOTICE THAT YOU CAN CANCEL TWO PAIRS OF ZEROS—ONE ZERO ON THE PERCENT SIGN WITH EACH ZERO ON THE NUMBERS!

10% and 1% as Decimal Slides

Examples

A. 10% of 37 = 3.7

REMEMBER THAT THERE IS AN INVISIBLE DECIMAL PLACE TO THE RIGHT OF THE ONE'S PLACE!

10% MEANS "DIVIDE BY 10!" THIS IS THE SAME AS SLIDING THE DECIMAL ONE TIME TO THE LEFT!

B. 1% of 37 = .37

1% MEANS "DIVIDE BY 100!" THIS IS THE SAME AS SLIDING THE DECIMAL ***TWO TIMES*** TO THE ***LEFT***!

Calculate the percentages of the following numbers

1. 10% of 58 = ☐
2. 1% of 63 = ☐
3. 10% of 30 = ☐
4. 50% of 18 = ☐
5. 25% of 28 = ☐
6. 20% of 50 = ☐
7. 10% of 50 = ☐
8. 5% of 20 = ☐
9. 10% of 72 = ☐
10. 1% of 358 = ☐
11. 5% of 40 = ☐
12. 1% of 25 = ☐
13. 10% of 70 = ☐
14. 1% of 70 = ☐
15. 10% of 47 = ☐
16. 1% of 47 = ☐
17. 50% of 14 = ☐
18. 20% of 15 = ☐
19. 10% of 248 = ☐
20. 5% of 100 = ☐

Percent Practice

Calculate the percentages of the following numbers

1. $33.\overline{3}\%$ of 60 = ☐
2. 1% of 500 = ☐
3. 1% of 365 = ☐
4. 50% of 28 = ☐
5. 20% of 45 = ☐
6. $33.\overline{3}\%$ of 15 = ☐
7. $66.\overline{6}\%$ of 15 = ☐
8. 10% of 80 = ☐

   \+ \+

9. 5% of 80 = ☐

   = =

10. 15% of 80 = ☐

15% IS JUST 10% + 5%!

11. 75% of 40 = ☐
12. 10% of 120 = ☐
13. 20% of 30 = ☐
14. 25% of 32 = ☐
15. $33.\overline{3}\%$ of 27 = ☐
16. 1% of 53 = ☐
17. $66.\overline{6}\%$ of 60 = ☐
18. 20% of 25 = ☐
19. 1% of 30 = ☐
20. 4% of 30 = ☐

4% IS JUST 4 TIMES WHAT 1% IS!

Concept Quiz

1. A percent is an amount out of ____________________

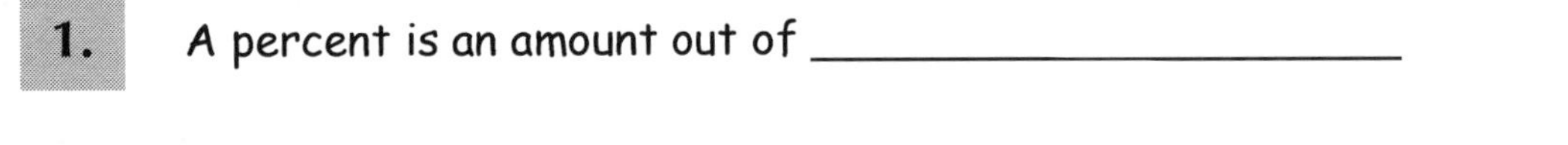

2. Break down the word "percent": per = "__________"; cent = "___________________"

3. Shade in the correct amount and fill in the percentage not shaded:

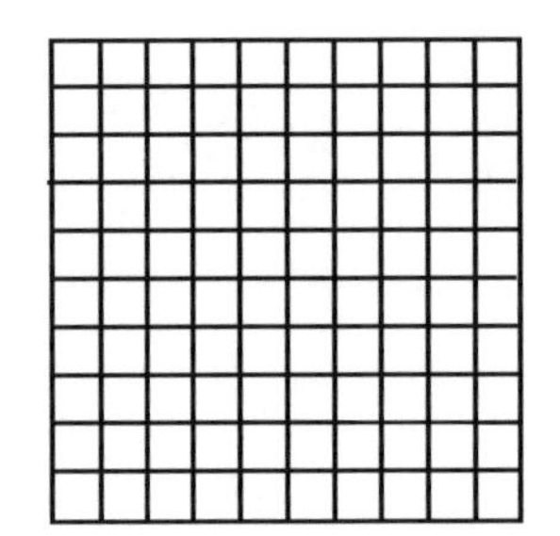

Shaded: 47 %

Not shaded: ____ %

4. If you can't remember what you divide by to get 25% of a number, what can you do to figure it out?

____

5. What is the easy trick to figure out 10% of a number?

____

6. What is the easy trick to figure out 1% of a number?

____

7. If you know 1% of a number, how could you easily calculate 2% of the same number?

____

The DP Trick

DR. DECIMAL

2 slides to the **LEFT**, if you have a PERCENT and want to make a DECIMAL

It takes 2 slides

← → ← →

to get from one side to another!

PEPPER PERCENT

2 slides to the **RIGHT**, if you have a DECIMAL and want to make a PERCENT

Use the DP Trick to change percents to decimals and the other way around

	DR. DECIMAL		PEPPER PERCENT
A.	.30 (or .3)	= ← ← slide the decimal point 2 times to the LEFT	30% (30.%)
1.	☐	← ← slide the decimal point 2 times to the LEFT	70%
2.	☐	← ← slide the decimal point 2 times to the LEFT	23%
3.	☐	← ← slide the decimal point 2 times to the LEFT	6% (06.%)
4.	☐	← ← slide the decimal point 2 times to the LEFT	245%

YOU NEED TO ADD A ZERO BEFORE THE 6 TO HAVE ROOM TO SLIDE TWO TIMES!

	Decimal		Percent
5.	.60	→ → slide the decimal point 2 times to the RIGHT	☐ %
6.	.37	→ → slide the decimal point 2 times to the RIGHT	☐ %
7.	.8 (= .80)	→ → slide the decimal point 2 times to the RIGHT	☐ %
8.	.75	→ → slide the decimal point 2 times to the RIGHT	☐ %
9.	3.25	→ → slide the decimal point 2 times to the RIGHT	☐ %

The DP Trick

Just slide two times to go from decimals to percents (or percents to decimals)

Use the DP Trick to change percents to decimals or the other way around

	DR. DECIMAL		PEPPER PERCENT					
1.	.26	=	☐ %		**10.**	5.7	=	☐ %
2.	☐	=	73%		**11.**	☐	=	.6%
3.	☐	=	80%		**12.**	☐	=	321%
4.	.4	=	☐ %		**13.**	.002	=	☐ %
5.	.05	=	☐ %		**14.**	36	=	☐ %
6.	9	=	☐ %		**15.**	☐	=	12%
7.	☐	=	247%		**16.**	☐	=	27.3%
8.	☐	=	6%		**17.**	3.14	=	☐ %
9.	.045	=	☐ %		**18.**	☐	=	500%

Percents as Division & Multiplication

Give the percent equivalents

1. Divide by 2 = ☐ %

2. Divide by 4 = ☐ %

3. Divide by 5 = ☐ %

4. Divide by 10 = ☐ %

5. Divide by 20 = ☐ %

6. Divide by 100 = ☐ %

7. Divide by 1 = ☐ %

8. ÷ by 3 or $\cdot \frac{1}{3}$ = ☐ %

9. Times $\frac{2}{3}$ = ☐ %

10. Times $\frac{3}{4}$ = ☐ %

VIEWED A SLIGHTLY DIFFERENT WAY...

11. 18 ÷ 2 is the same as ☐ % of 18

12. 40 ÷ 10 is the same as ☐ % of 40

13. 35 ÷ 5 is the same as ☐ % of 35

14. 12 ÷ 3 is the same as ☐ % of 12

15. 32 ÷ 4 is the same as ☐ % of 32

16. 24 ÷ 3 · 2 is the same as ☐ % of 24

Percent Review

1.

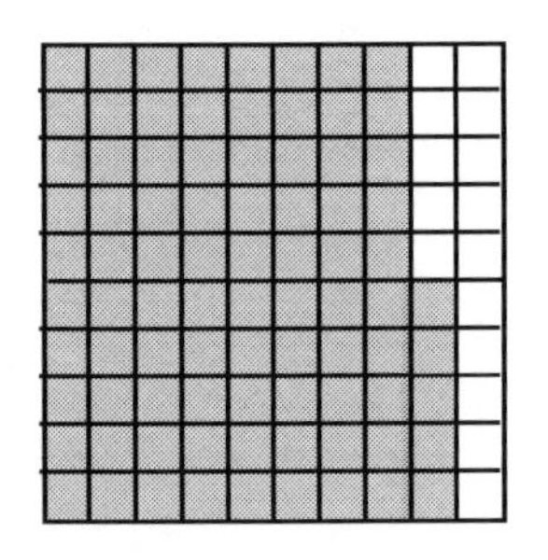

Shaded:

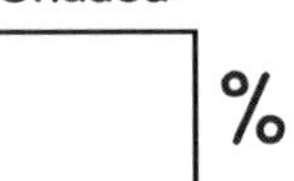 %

Not shaded:

☐ %

2. How many times does 20 go into 100? ☐

20% = Divide by ☐

or times $\frac{1}{☐}$

3. Divide by 4 = ☐ %

DR. DECIMAL		PEPPER PERCENT
4. .30	=	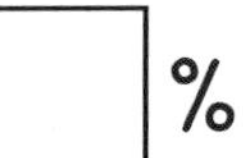 %
5. ☐	=	72%
6. .09	=	☐ %
7. ☐	=	8%
8. 2.5	=	☐ %

9. 25% of 40 = 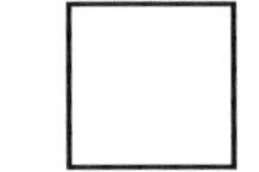

10. Times $\frac{1}{3}$ = ☐ %

11. Times $\frac{2}{3}$ = ☐ %

12. Times $\frac{3}{4}$ = ☐ %

13.

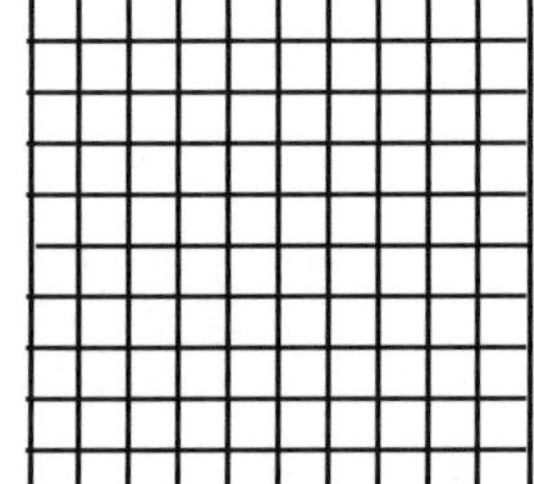

Shade 23%

Not shaded:

☐ %

14. Divide by 10 = ☐ %

15. 
Divide by 1 = ☐ %

16. 70% = 10% × ☐

17. How does 5% differ from 10%

Percent Practice

Calculate the percentages of the following numbers.

1. 50% of 70 =
2. 100% of 8 =
3. 150% of 8 =
4. 25% of 32 =
5. 25% of 12 =
6. 125% of 12 =
7. 10% of 90 =
8. 20% of 90 =
9. 30% of 90 =
10. 20% of 30 =
11. 20% of 45 =
12. 25% of 44 =
13. 10% of 46 =
14. 1% of 46 =
15. 10% of 60 =
16. 100% of 60 =
17. 110% of 60 =
18. 50% of 16 =
19. 25% of 16 =
20. 1% of 365 =

NOW TRY THESE PROBLEMS!

21. 10% of 30 =
22. 70% of 30 =
23. What is the price of a $40 meal plus a 15% tip?

Concept Quiz

1. Percents and decimals are just different ways of writing the same number: True / False
(circle one)

2. How many slides does it take to go from a decimal to a percent (or the other way around)?

____

3. Why does it take this number of slides? ______________________________

____

4. Which way do you slide if you have a percent and want to write a decimal?

Slide to the LEFT / Slide to the RIGHT (circle one)

5. Divide by 5 is the same as WHAT PERCENT of a number? __________

6. How can you figure this out if you need to do so?

____

7. If you know 10% of a number, how can you easily calculate 30% of the same number?

____

8. How does 5% differ from 10%?

____

Percents in Terms of Other Percents

Calculate, using basic operations with percents

1. 30% = 10% × ☐

2. 70% = 10% × ☐

3. 80% = 20% × ☐

4. 50% = 25% × ☐

5. 50% = 10% × ☐

6. 90% = 10% × ☐

7. 40% = 20% × ☐

8. 25% = 50% ÷ ☐

9. 5% = 50% ÷ ☐

10. 5% = 10% ÷ ☐

11. 100% = 20% × ☐

12. 20% = 10% × ☐

13. 60% = 10% × ☐

14. 60% = 20% × ☐

15. 15% = 10% + ☐ %

16. 25% = 20% + ☐ %

17. 15% = 20% - ☐ %

18. 90% = 100% - ☐ %

19. 80% = 100% - ☐ %

20. 30% = 20% + ☐ %

PUT THIS SKILL TO GOOD USE!

21. 35% of 50 =

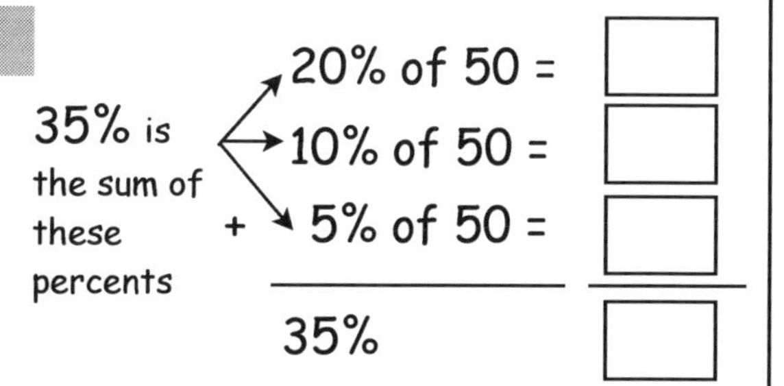

35% is the sum of these percents
20% of 50 =
10% of 50 =
+ 5% of 50 =
35%

Percent Review

1.

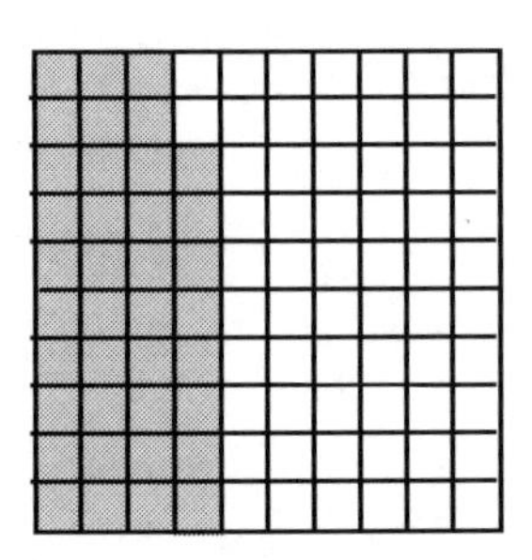

Shaded:

 %

Not shaded:

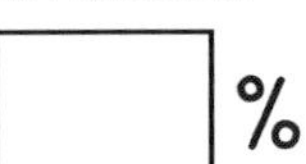 %

2. How many times does 5 go into 100? ☐

5% = Divide by ☐

or times $\frac{1}{\square}$

3. Divide by 100 = ☐ %

	DR. DECIMAL		PEPPER PERCENT
4.	.02	=	☐
5.		=	125%
6.	.57	=	☐ %
7.	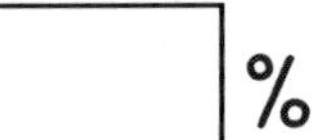	=	8%
8.	.006	=	☐ %

9. 5% of 40 = ☐

10. Times $\frac{3}{4}$ = ☐ %

11. Times $\frac{1}{3}$ = ☐ %

12. Times $\frac{2}{3}$ = ☐ %

13.

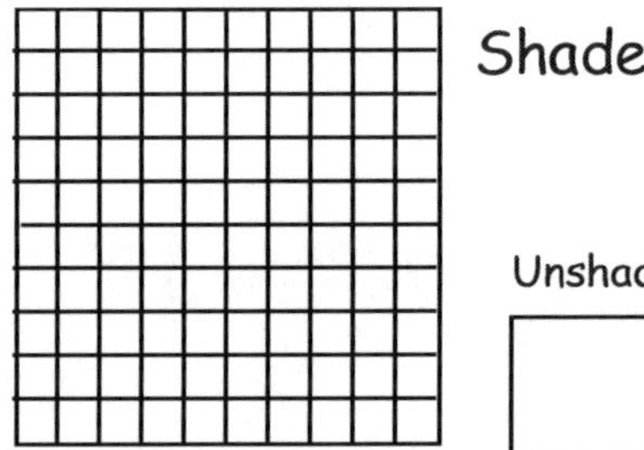

Shade 83%

Unshaded:

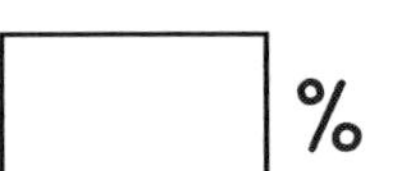 %

14. 40% = 10% × ☐

15. 60% = 20% × ☐

16. $66.\overline{6}\%$ = $33.\overline{3}\%$ × ☐

17. What is one way you could calculate 45%?

Using 10% as a Reference

Use 10% as a reference to figure out other percent problems

Example

A. 10% of 30 = → 3
× 4 × 4
40 % of 30 = 12

1. 10% of 70 = → ☐
× 3 × 3
☐ % of 70 = ☐

2. 10% of 60 = → ☐
× 9 × 9
☐ % of 60 = ☐

3. 10% of 50 = → ☐
× 7 × 7
☐ % of 50 = ☐

4. 10% of 120 = → ☐
× 8 × 8
☐ % of 120 = ☐

5. 10% of 20 = → ☐
× 6 × 6
☐ % of 20 = ☐

6. 10% of 40 = → ☐
× 7 × 7
☐ % of 40 = ☐

7. 10% of 80 = → ☐
× 6 × 6
☐ % of 80 = ☐

8. 10% of 50 = → ☐
÷ 2 ÷ 2
☐ % of 50 = ☐

9. 10% of 30 = → ☐
÷ 2 ÷ 2
☐ % of 30 = ☐

Percent Practice

Calculate the percentages of the following numbers

1. 10% of 45 = ☐
2. 1% of 45 = ☐
3. $33\frac{1}{3}\%$ of 27 = ☐
4. $66\frac{2}{3}\%$ of 27 = ☐
5. $66.\overline{6}\%$ of 18 = ☐
6. $33.\overline{3}\%$ of 18 = ☐
7. 75% of 44 = ☐
8. 10% of 40 = ☐
   \+ \+
9. 5% of 40 = ☐
   = =
10. 15% of 40 = ☐

15% IS JUST 10% + 5%!

11. 20% of 50 = ☐
12. $66\frac{2}{3}\%$ of 24 = ☐
13. 20% of 35 = ☐
14. 75% of 36 = ☐
15. $33.\overline{3}\%$ of 60 = ☐
16. 300% of 7 = ☐
17. $66.\overline{6}\%$ of 90 = ☐
18. 1% of 500 = ☐
19. 2% of 500 = ☐
20. 6% of 500 = ☐

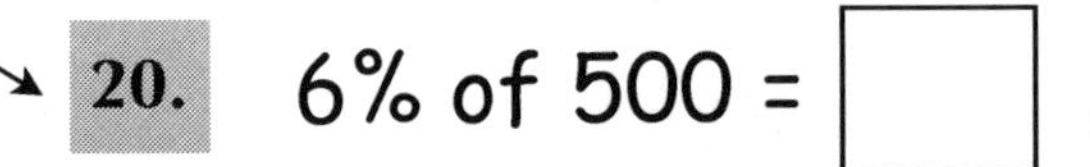

COMPARE THESE TO 1%!

12.5% and Other Fraction Percents

Example

A. $25\% = \frac{1}{4}$ or divide by 4

↓ cut in half ↓ ↓

$12.5\% = \frac{1}{8}$ or divide by 8 $12\frac{1}{2}\%$ of 24 = 3

THESE NUMBERS ARE ALL HALF OF THE TOP NUMBERS!

12.5% IS THE SAME AS 12 1/2%!

Calculate the percentages of the following numbers

1. 12.5% of 40 = ☐

2. $33\frac{1}{3}\%$ of 60 = ☐

3. 75% of 28 = ☐

4. 12.5% of 32 = ☐

5. $66\frac{2}{3}\%$ of 33 = ☐

6. $33\frac{1}{3}\%$ of 21 = ☐

7. $12\frac{1}{2}\%$ of 72 = ☐

8. 700% of 6 = ☐

9. $66\frac{2}{3}\%$ of 30 = ☐

10. 12.5% of 64 = ☐

REVIEW PROBLEMS

11. 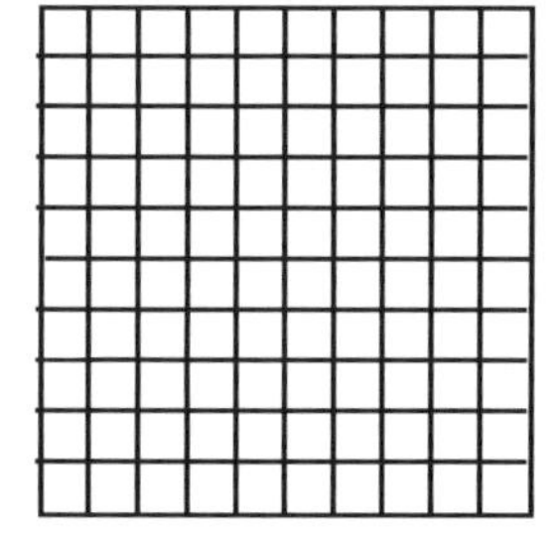 Shade 49%

Unshaded: ☐ %

12. 90% = 10% × ☐

13. Divide by 4 = ☐ %

14. Divide by 8 = ☐ %

Percent Practice

Calculate the percentages of the following numbers

1. 20% of 40 =
2. 100% of 60 =
3. $12\frac{1}{2}$% of 16 =
4. $33\frac{1}{3}$% of 30 =
5. 10% of 45 =
6. 1% of 16 =
7. $66\frac{2}{3}$% of 18 =
8. 300% of 20 =
9. $12\frac{1}{2}$% of 24 =
10. 50% of 48 =
11. $66.\overline{6}$% of 21 =
12. 25% of 80 =
13. 5% of 80 =

ANOTHER WAY TO CALCULATE THIS ONE IS TO REMEMBER THAT 5% IS FIVE TIMES SMALLER THAN 25%!

14. 10% of 6 =
15. 30% of 6 =

JUST 3 TIMES WHAT 10% OF 6 IS!

16. 75% of 48 =
17. 20% of 50 =
18. 1% of 2000 =
19. 3% of 2000 =
20. 6% of 2000 =

COMPARE THESE LAST TWO TO 1%!

100% and Above

Example

A. 1.00 = 100%

Two zeros on the percent sign = Two slides on the decimal

In other words 100% is the same as the number 1

Write these amounts for percents that are greater than 1

1. 200% =

2. 300% =

3. 700% =

4. 1000% =

5. 800% =

6. 12,000% =

7. 400% =

8. 600% =

9. 900% =

A LITTLE HARDER...

10. 200% of 9 = 

REMEMBER THAT "OF" MEANS "TIMES"

11. 300% of 7 =

12. 500% of 9 =

13. 800% of 2 =

13. 700% of 4 = 

14. 600% of 3 =

15. 100% of 9 =

16. 1200% of 3 =

17. 1100% of 6 =

Percents by Making Denominators a Hundred

WHEN THE BOTTOM NUMBER IS 100,
THE TOP NUMBER TELLS THE PERCENT!

Example

A. 3 out of 50 is what percent? 6 %

#1 Ask yourself, "Can I multiply the bottom number by any whole number to get 100?"

#2 If you can, multiply the top number by the same amount—and this number automatically tells the percent!

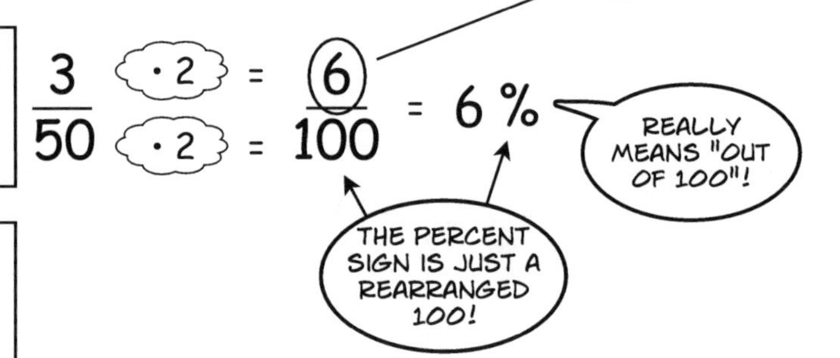

$\frac{3}{50}$ ·2 ·2 = $\frac{6}{100}$ = 6 %
REALLY MEANS "OUT OF 100"!
THE PERCENT SIGN IS JUST A REARRANGED 100!

Figure out the percent by making the bottom number equal to 100

1. 7 out of 10 is what percent?

$\frac{7}{10}$ (·) (·) = $\frac{\square}{100}$ $\square$ %

2. What percent is 4 out of 5?

$\frac{\square}{\square}$ (·) (·) = $\frac{\square}{100}$ $\square$ %

3. 15 out of 20 players on the team got base hits. What percent was that?

$\frac{\square}{\square}$ (·) (·) = $\frac{\square}{100}$ $\square$ %

4. Twenty-five students are in Mr. Smith's class. Three of them were absent yesterday. What percent was that?

$\frac{\square}{\square}$ (·) (·) = $\frac{\square}{\square}$ $\square$ %

5. Five friends went out to eat. Three ordered cheeseburgers. What percent was this?

$\frac{\square}{\square}$ (·) (·) = $\frac{\square}{\square}$ $\square$ %

6. 14 out of 200 animals at the zoo bite the zookeepers regularly. What percent do this?

$\frac{\square}{\square}$ (÷) (÷) = $\frac{\square}{\square}$ $\square$ %

If the Bottom Number is 100

---THE TOP NUMBER TELLS THE PERCENT!

Example

A. $\frac{3}{50} = \frac{\boxed{6}}{100} = \boxed{6}\,\%$

A PERCENT IS A REARRANGED 100!

same thing!

Calculate the percentages of the following numbers

1. $\frac{10}{25} = \frac{\square}{100} = \square\,\%$

same thing!

2. $\frac{3}{20} = \frac{\square}{100} = \square\,\%$

3. $\frac{1}{4} = \frac{\square}{100} = \square\,\%$

4. $\frac{2}{4} = \frac{\square}{100} = \square\,\%$

5. $\frac{3}{4} = \frac{\square}{100} = \square\,\%$

6. $\frac{1}{5} = \frac{\square}{100} = \square\,\%$

7. $\frac{3}{5} = \frac{\square}{100} = \square\,\%$

8. $\frac{2}{10} = \frac{\square}{100} = \square\,\%$

9. 40 out of 50

out of $\frac{\square}{\square} = \frac{\square}{100} = \square\,\%$

10. 2 out of 5

out of $\frac{\square}{\square} = \frac{\square}{100} = \square\,\%$

11. 7 out of 25

out of $\frac{\square}{\square} = \frac{\square}{100} = \square\,\%$

Percents as Fractions

Example

A. $75\% = \frac{75}{100}$ — same thing, really

WHEN THE BOTTOM NUMBER'S A HUNDRED, THE TOP NUMBER TELLS THE PERCENT!

$\frac{75 \div 25}{100 \div 25} = \frac{3}{4}$

ALWAYS TRY TO SIMPLIFY YOUR ANSWER!

Convert the following percents to fractions

1. $20\% = \frac{\square}{100} = \frac{\div 20}{\div 20}\ \frac{\square}{\square}$

2. $30\% = \frac{\square}{100} = \frac{\div \ }{\div \ }\ \frac{\square}{\square}$

3. $50\% = \frac{\square}{\square} = \frac{\div \ }{\div \ }\ \frac{\square}{\square}$

4. $25\% = \frac{\square}{\square} = \frac{\div \ }{\div \ }\ \frac{\square}{\square}$

5. $10\% = \frac{\square}{\square} = \frac{\div \ }{\div \ }\ \frac{\square}{\square}$

6. $15\% = \frac{\square}{\square} = \frac{\div \ }{\div \ }\ \frac{\square}{\square}$

7. $35\% = \frac{\square}{\square} = \frac{\div \ }{\div \ }\ \frac{\square}{\square}$

8. $60\% = \frac{\square}{\square} = \frac{\div \ }{\div \ }\ \frac{\square}{\square}$

9. $40\% = \frac{\square}{\square} = \frac{\div \ }{\div \ }\ \frac{\square}{\square}$

10. $5\% = \frac{\square}{\square} = \frac{\div \ }{\div \ }\ \frac{\square}{\square}$

write in simplest form

11. $120\% = \frac{\square}{\square} = \square\ \frac{\square}{\square}$

write in simplest form

12. $250\% = \frac{\square}{\square} = \square\ \frac{\square}{\square}$

Percent Review

1.

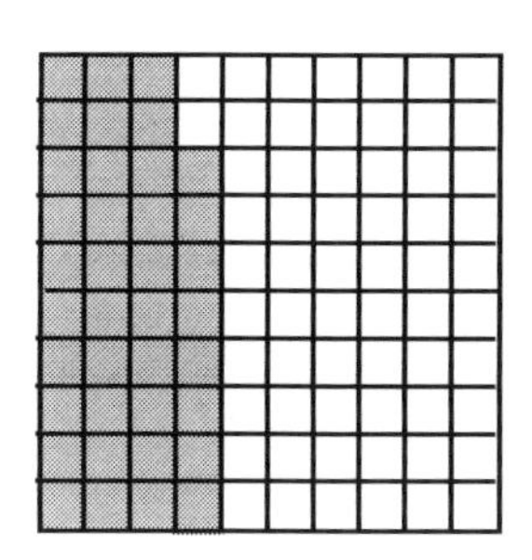

Shaded:

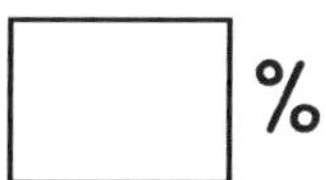 %

Not shaded:

 %

2. 12.5% of 48 = ☐

3. Divide by 25 = ☐ %

4. $\frac{7}{20} = \frac{\square}{100} = \square$ %

5. 60% = $\frac{\square}{100} = \frac{\square}{\square}$

	DR. DECIMAL		PEPPER PERCENT
6.	.02	=	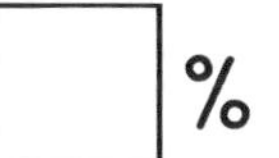 %
7.	☐	=	125%
8.	.57	=	☐ %
9.	☐	=	8%
10.	.006	=	☐ %

11. 15 out of 50

$\frac{\square}{\square} = \frac{\square}{100} = \square$ %

12.

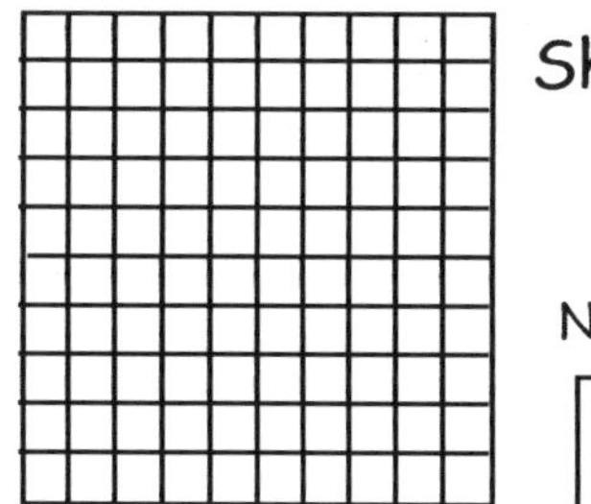

Shade 56%

Not shaded:

☐ %

13. What percent is 2 out of 5?

$\frac{\square}{\square}$ (•) = $\frac{\square}{100}$ ☐ %

14. 3 out of 25

$\frac{\square}{\square} = \frac{\square}{100} = \square$ %

15. 75% = 25% × ☐

16. 60% = 50% + ☐ %

17. What is one way you could calculate 90%?

☐

Working with Percents

Calculate the percentages of the following numbers

1. 50% of 12 = ☐
2. 150% of 12 = ☐
3. 10% of 50 = ☐
4. 5% of 50 = ☐
5. 15% of 50 = ☐
6. 20% of 35 = ☐
7. 75% of 28 = ☐
8. $33\frac{1}{3}$% of 60 = ☐
9. $66\frac{2}{3}$% of 12 = ☐
10. 25% of 36 = ☐

Express the division problems as percents

11. Divide by 5 = ☐ %
12. Divide by 10 = ☐ %
13. Divide by 100 = ☐ %
14. Divide by 2 = ☐ %
15. Divide by 3 = ☐ %
16. Divide by 3 and times by 2 = ☐ %
17. Divide by 4 = ☐ %
18. Divide by 4 and times by 3 = ☐ %

Percent Practice

Solve the following percent problems

1. 50% of 16 = ☐
2. 25% of 16 = ☐
3. 75% of 16 = ☐
4. 10% of 23 = ☐
5. 10% of 10 = ☐
6. 20% of 10 = ☐
7. 10% of 120 = ☐

HALF OF 10%

8. 5% of 120 = ☐

SAME AS 10% + 5%!

9. 15% of 120 = ☐
10. 20% of 45 = ☐

11. 50% is ÷ by ☐
12. 25% is ÷ by ☐
13. 20% is ÷ by ☐
14. 10% is ÷ by ☐
15. 5% is ÷ by ☐
16. 33.3% is ÷ by ☐
17. 66.6% is ÷ by and • by

Bottom of a fraction = divide
Top of a fraction = multiply

☐ ☐

18. 75% is ÷ by and • by

☐ ☐

19. 100% is ÷ by ☐
20. 300% is • by ☐

DP Trick Practice

Just slide two times to go from decimals to percents (or percents to decimals)

Use the DP Trick to change percents to decimals or the other way around

1. Write 8% as a decimal ☐

2. Write 23% as a decimal ☐

3. Write .3 as a percent ☐ %

4. Write .42 as a percent ☐ %

5. Write 20% as a decimal ☐

6. Write 200% as a decimal ☐

7. Write .9 as a percent ☐ %

8. Write 9 as a percent ☐ %

9. Write .4% as a decimal ☐

10. Write 8.2% as a decimal ☐

11. Tommy tried to impress people with his age by telling people that his age was 1600%. How old is he really?

☐

12. Sally tried to confuse her friends by saying that she paid 75% of a dollar for a candy bar. How much was it really?

☐

13. What is 15% as a decimal?

☐

14. At a certain electronics store the prices increased by a factor of .25. What is that factor as a percent?

☐

15% Restaurant Tips

Example

A. Calculate the cost of a $20 meal with a 15% tip

Cost of meal:	$20	
15% tip: +	10% of $20: $2	cut this in half to get 5%
	5% of $20: $1	add the 10% and the 5% to get 15%
	15% $3	

Cost of meal plus the tip: $23

Calculate the cost of the meal PLUS A 15% TIP

1. Calculate the cost of a $60 meal with a 15% tip

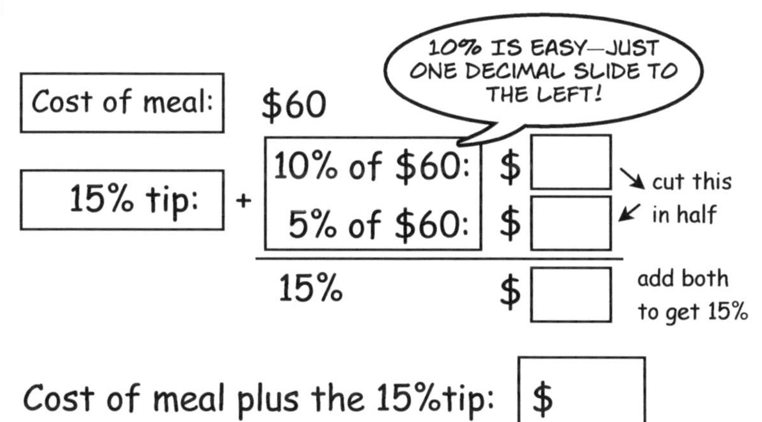

Cost of meal plus the 15%tip: $

2. Calculate the cost of a $50 meal with a 15% tip

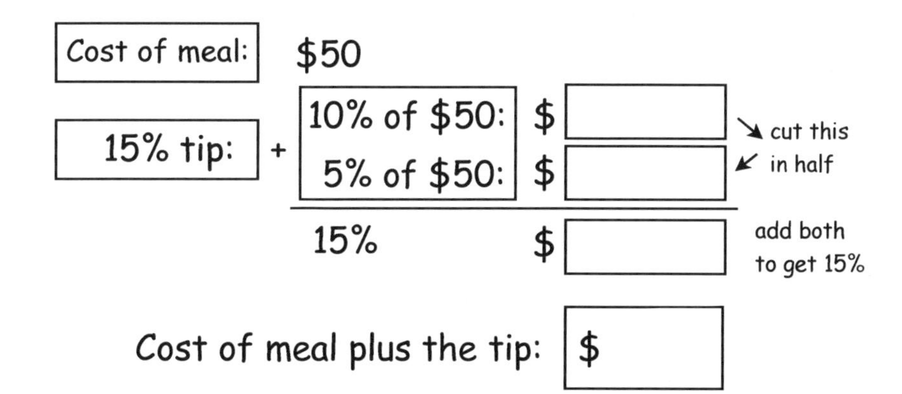

Cost of meal: $50
15% tip: +
10% of $50: $
5% of $50: $
cut this in half
15% $
add both to get 15%
Cost of meal plus the tip: $

15% Restaurant Tip Practice

Calculate the cost of the meal plus a 15% tip

1. Calculate the cost of a $140 meal with a 15% tip

Cost of meal: $140

15% tip: +
10% of $140: $ ☐ ↘ cut this
5% of $140: $ ☐ ↙ in half

15% $ ☐ add both to get 15%

Cost of meal plus the 15% tip: $ ☐

2. Calculate the cost of a $100 meal with a 15% tip

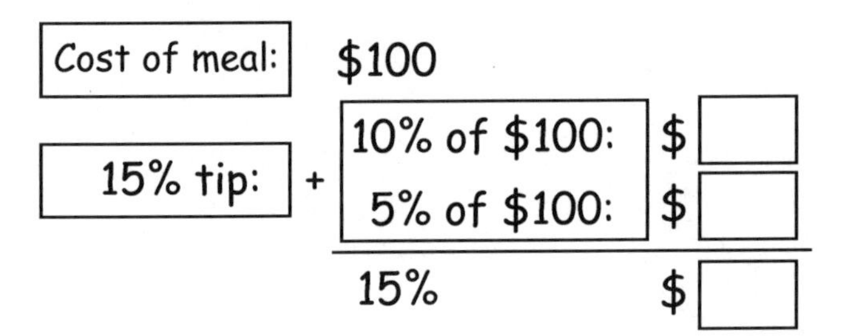

Cost of meal plus the 15% tip: $ ☐

3. Calculate the cost of a $30 meal with a 15% tip

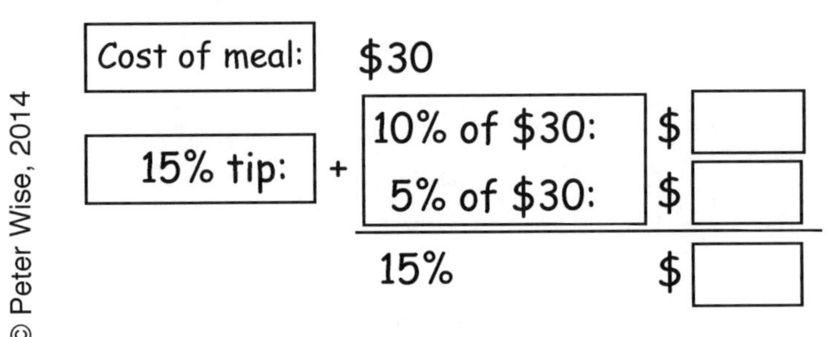

Cost of meal:
$30
15% tip:
+
10% of $30:
$
5% of $30:
$
15%
$

Cost of meal plus the 15%tip: $ ☐

4. Calculate the cost of a $12 meal with a 15% tip

Cost of meal: $12

15% tip: +
10% of $12: $ ☐
5% of $12: $ ☐

15% $ ☐

Cost of meal plus the 15% tip: $ ☐

5. Calculate the cost of a $180 meal with a 15% tip

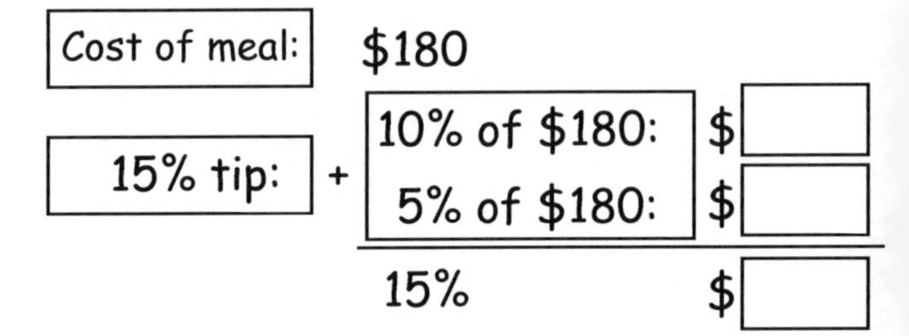

Cost of meal:
$180
15% tip:
+
10% of $180:
$
5% of $180:
$
15%
$

Cost of meal plus the 15% tip: $ ☐

6. Calculate the cost of a $24 meal with a 15% tip

Cost of meal: $24

15% tip: +
10% of $24: $ ☐
5% of $24: $ ☐

15% $ ☐

Cost of meal plus the 15% tip: $ ☐

Price With 10% Discount

Example

A. Calculate the cost of a $30 book with a 10% discount

Cost of item: $30

\- 10% of $30: $3

$27

10% IS EASY TO FIGURE OUT—JUST TAKE OFF A ZERO OR SLIDE ONE TIME TO THE LEFT!

DISCOUNTS ARE ALWAYS AMOUNTS SUBTRACTED!

Discounted amounts are always lower than the original amount

Calculate the cost of the following items with a 10% discount

1. Calculate the cost of a $20 book with a 10% discount

Cost of item:

\- 10%:

$

2. Calculate the cost of a $40 pair of shoes with a 10% discount

Cost of item:

\- 10%:

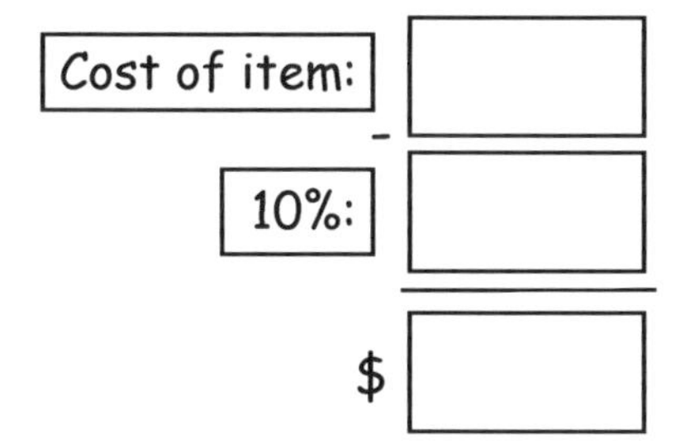

$

3. Calculate the cost of a $60 jacket with a 10% discount

Cost of item:

\- 10%:

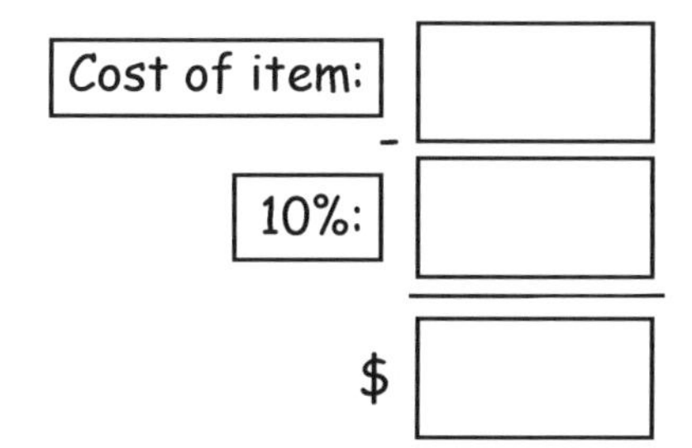

$

4. Calculate the cost of a $35 sweater with a 10% discount

Cost of item:

\- 10%:

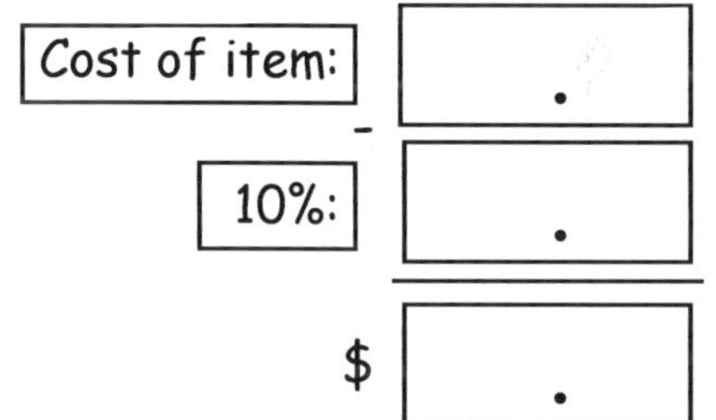

$

5. Calculate the cost of a $48 gift with a 10% discount

Cost of item:

\- 10%:

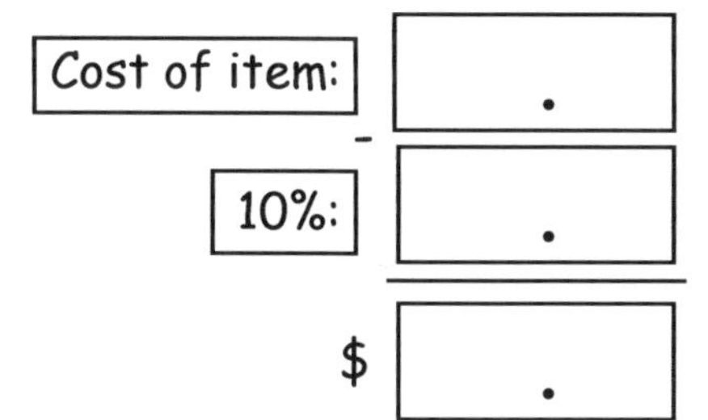

$

6. Calculate the cost of a $150 gift with a 10% discount

Cost of item:

\- 10%:

$

Cost with 10% Tax

Example

A. Calculate the cost of a $20 book with 10% tax

Cost of item:	$20
+ 10% of $20:	$2
	$22

TAXES ARE ALWAYS ADDED!

Taxes are always added to the original amount

Calculate the cost of the following items with a 10% tax

1. Calculate the cost of a $80 jacket with a 10% tax

Cost of item: ☐
+ 10%: ☐
Cost with tax: ☐

2. Calculate the cost of a $30 pair of shoes with a 10% tax

Cost of item: ☐
+ 10%: ☐
Cost with tax: ☐

3. Calculate the cost of a $60 meal with a 10% tax

Cost of item: ☐
+ 10%: ☐
Cost with tax: ☐

4. Calculate the cost of a $15 shirt with a 10% tax

Cost of item: ☐
+ 10%: ☐
Cost with tax: ☐

5. Calculate the cost of a $52 pair of shoes with 10% tax

Cost of item: ☐
+ 10%: ☐
Cost with tax: ☐

6. Calculate the cost of a $12.50 belt with 10% tax

Cost of item: ☐
+ 10%: ☐
Cost with tax: ☐

Concept Quiz

1. This percent is the same as the whole number ONE: ______________

2. This percent is the same as the whole number FIVE: ______________

3. The top number of a fraction automatically tells you _______ ______________if the bottom number is ______________________ .

5. After figuring out a TIP, you take the original amount and then ADD / SUBTRACT a percent amount. (circle one)

4. It's common to pay a 15% tip at a restaurant. What is an easy way to calculate 15% of $60? Calculate the tip and the final amount.

Step 1: ____

Step 2: ____

Step 3: ____

Step 4 (final amount): ____________________________________

6. When figuring out a DISCOUNT, you take the original cost and then ADD / SUBTRACT a percent amount. (circle one)

7. How would you calculate the price of a $40 item at the store with a 10% discount?

Step 1: ____

Step 2 (final amount): ____________________________________

What is a% of b?

Example

A. What is 32% of 96?

#1 Convert 32% to a decimal (use the DP Trick)

32% = .32

#2 "of" means TIMES ...so just multiply the decimal times the other number

$$\begin{array}{r} 96 \\ \times .32 \\ \hline \end{array}$$

Take the decimal off
• Multiply normally;
• Put the decimal back on at the end!

$$\begin{array}{r} 96 \\ \times\ 32 \\ \hline 192 \\ 2880 \\ \hline 3072 \end{array} \rightarrow \boxed{30.72}$$

THIS IS THE ANSWER!

Now put on the decimal by sliding two times LEFT

Shortcut! You can also just multiply the percent times the other number and just put two slides on your answer

Calculate the following by converting the percents to decimals and multiplying

1. What is 14% of 82?

#1 Convert the percent to a decimal (use the DP Trick)

#2 Multiply this new decimal times 82

Answer:

2. What is 35% of 75?

#1 Decimal form of the percent:

#2 Multiply the numbers now:

Answer:

3. What is 57% of 63?

Percent as a decimal:

Answer:

4. What is 72% of 45?

Percent as a decimal:

Answer:

Practice with a% of b

ANOTHER WAY TO CALCULATE PERCENTS...
SOME PEOPLE FIND THIS METHOD EASIER

#1 Multiply the percent amount and the number normally

#2 Now give your answer two slides LEFT (because the % has 2 zeros!)

Example

A. What is 42% of 67?

Multiply normally: 42 x 67 = 2,814

Slide your answer to the LEFT 2 times: 26.14

Multiply normally and give your answer two slides

1. What is 35% of 40?

Multiply normally:

Give your answer 2 slides:

2. What is 47% of 90?

Multiply normally:

Give your answer 2 slides:

3. What is 18% of 75?

Multiply normally:

Give your answer 2 slides:

4. What is 23% of 68?

Multiply normally:

Give your answer 2 slides:

5. What is 8% of 36?

Multiply normally:

Give your answer 2 slides:

6. What is 3% of 120?

Multiply normally:

Give your answer 2 slides:

Practice with a% of b

Multiply normally and give your answer two slides

1. What is 20% of 35?

2. What is 63% of 75? 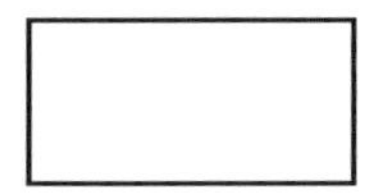

3. What is 92% of 80? 

4. What is 48% of 36?

5. What is 18% of 54?

6. What is 2% of 87? 

7. What is 14% of 96? 

8. What is 3% of 120?

9. What is 170% of 38?

10. What is 321% of 74?

Different Rates of Tax

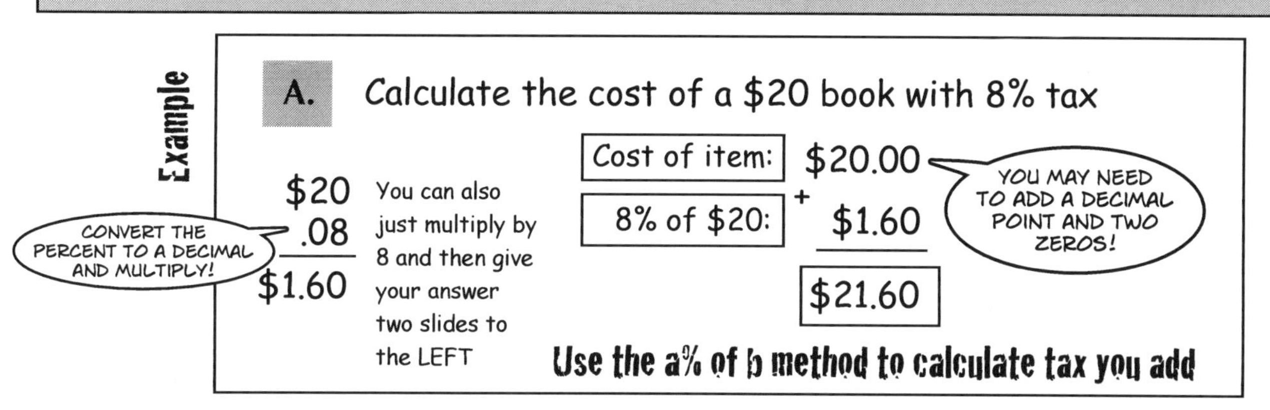
Example
A.
Calculate the cost of a $20 book with 8% tax
$20
.08
$1.60
CONVERT THE PERCENT TO A DECIMAL AND MULTIPLY!
You can also just multiply by 8 and then give your answer two slides to the LEFT
Cost of item: $20.00
8% of $20: + $1.60
$21.60
YOU MAY NEED TO ADD A DECIMAL POINT AND TWO ZEROS!
Use the a% of b method to calculate tax you add

Calculate the cost of the following items with different tax rates

1. Calculate the cost of a $10 binder with 7% tax

Cost of item:
+
Cost times tax rate:

2. Calculate the cost of a $30 photo frame plus 8% tax

Cost of item:
+
Cost times tax rate:

3. Calculate the cost of a $80 jacket plus 9% tax

Cost of item:
+
Cost times tax rate:

4. Calculate the cost of a $12 shirt with 7% tax

Cost of item:
+
Cost times tax rate:

5. Calculate the cost of a $15 meal with 8% tax

Cost of item:
+
Cost times tax rate:

6. Calculate the cost of a $35 park pass with 6% tax

Cost of item:
+
Cost times tax rate:

Finding Totals and Tax

Calculate the cost of the following items with different tax amounts

1. Randy bought two adventure books at $12.50 each and four biographies at $8.50 each. The tax rate was 7%. Calculate the total cost, including tax. How much change would his parents get if they paid with a $100 bill?

$12.50 x ☐ = ☐
+
$8.50 x ☐ = ☐

SUBTOTAL ☐
+
SUBTOTAL x 7% tax = ☐

TOTAL ☐

hundred dollars ☐
- ☐

CHANGE ☐

2. Suzie bought four bags of pet food for $3.10 each and 6 plastic toys for $2.80 each. The tax rate in her city is 6%. Find her total amount.

☐ x ☐ = ☐
+
☐ x ☐ = ☐

SUBTOTAL ☐
+
SUBTOTAL x 6% tax = ☐ Round your answer to the nearest 100th

TOTAL ☐

Finding Totals and Tax

Calculate the cost of the following items with a 10% tax

1. Sammy bought two shirts at $4.50 each and three pair of pants at $10 each. What was the cost of all the clothes and a 10% tax?

Two shirts at $4.50 each [ ]

+

Three pairs of pants at $10 each [ ]

SUBTOTAL [ ]

+

10% [ ]

TOTAL [ ]

2. Cindy's mom took the family out to lunch. She bought four hamburgers at $3.25 each and three orders of fries at $1.50 each. How much was the total if there was a 10% tax?

hamburgers [ ]

+

fries [ ]

[ ]

+

tax amount [ ]

TOTAL [ ]

Percent Review

1. Shade 45%

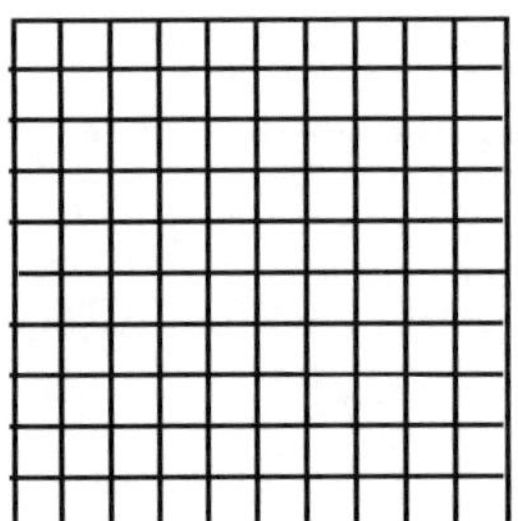

Unshaded: ☐ %

2. Divide by 8 = ☐ %

3. $\frac{12}{25} = \frac{\square}{100} = \square\ \%$

4. 35% =

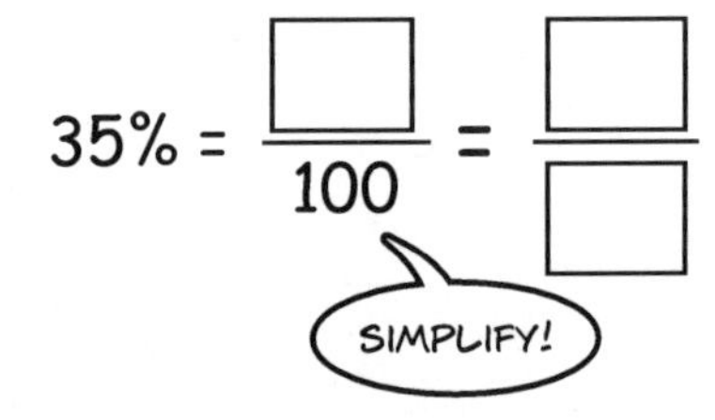

5. Calculate the cost of a \$70 jacket with a 10% discount ☐

	DECIMAL		PERCENT
6.	.02	=	☐ %
7.	☐	=	125%
8.	.57	=	☐ %
9.	☐	=	8%

10. 15 out of 50

out of $\frac{\square}{\square} = \frac{\square}{100} = \square\ \%$

11. Calculate the cost of a \$50 meal with a 15% tip

☐

12. What percent is 15 out of 20?

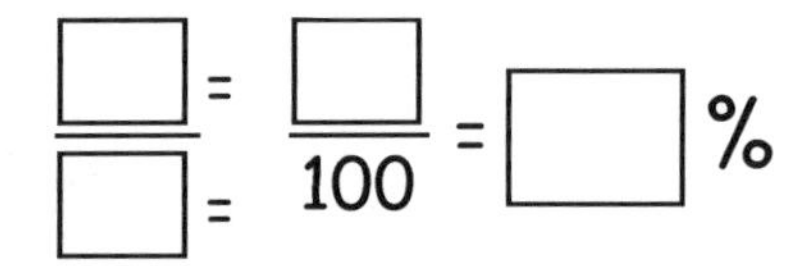

13. Calculate the cost of a \$12 binder with 6% tax

☐

14. Katie bought 3 bags of pet food for \$4.00 each and 6 plastic toys for \$3.00 each. The tax rate in her city is 6%. Find her total amount.

☐

Percents Test

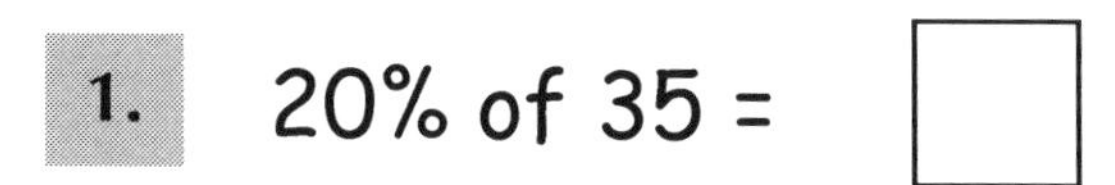

1. 20% of 35 = ☐
2. 25% of 24 = ☐
3. 10% of 70 = ☐
4. 5% of 80 = ☐
5. 50% of 36 = ☐
6. $66\frac{2}{3}$% of 12 = ☐
7. 200% of 8 = ☐
8. 75% of 28 = ☐
9. $33\frac{1}{3}$% of 27 = ☐
10. 10% of 43 = ☐

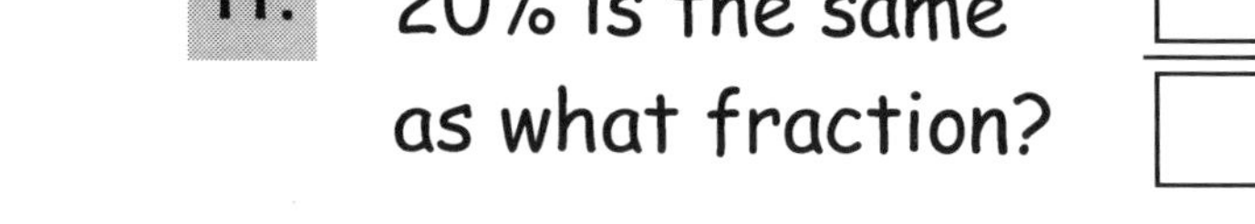

PUT IN SIMPLEST FORM!

11. 20% is the same as what fraction? $\frac{☐}{☐}$
12. 1% of 17 = ☐
13. 150% of 20 = ☐
14. 70% of 40 = ☐
15. 15% of 80 = ☐
16. 60% = what decimal? ☐
17. 7% = what decimal? ☐
18. 25% of 16 = ☐
19. 5% of 40 = ☐
20. 10% of .4 = ☐

Advanced Percents

- Finding the Percent
- Finding the Whole
- Solving for Percents
- Using the Algebraic Method

Find the Percent

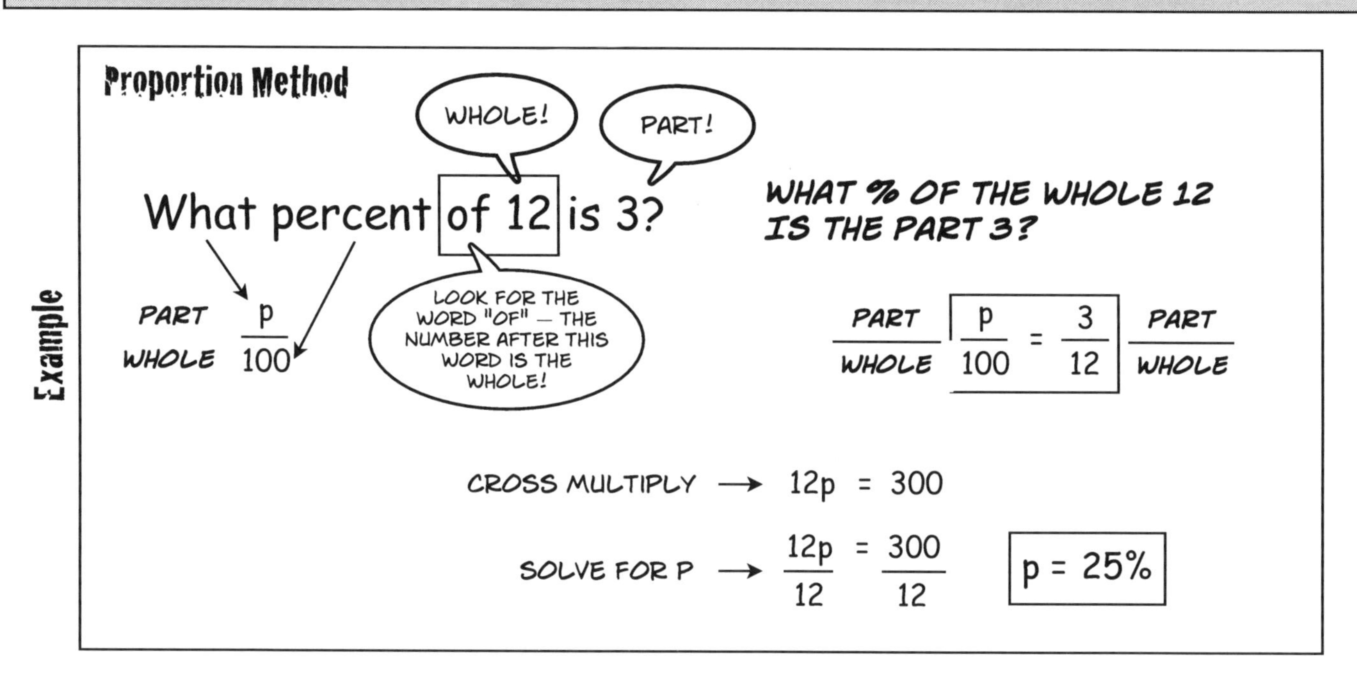

Find the PERCENT, using RATIOS (a.k.a. PROPORTIONS)

1. What percent of 40 is 6? Answer: ☐ %

Remember!
Whenever you see "WHAT PERCENT" think... → $\frac{p}{100}$ = $\frac{\square}{\square}$ ← PART ← WHOLE

THE 6 GOES IN THE LAST PLACE AVAILABLE!

THE NUMBER AFTER THE WORD "OF" GOES ON THE BOTTOM!

2. What percent of 12 is 5? Answer: ☐ %

show the remainder as a fraction

$\frac{p}{\square}$ = $\frac{\square}{\square}$ ← PART ← WHOLE

3. What percent of 5 is 27? Answer: ☐ %

$\frac{\square}{\square}$ = $\frac{\square}{\square}$ ← PART ← WHOLE

Find the Percent

Find the PERCENT using PROPORTIONS; round to the nearest tenth

Answer:

1. What percent of 35 is 7? 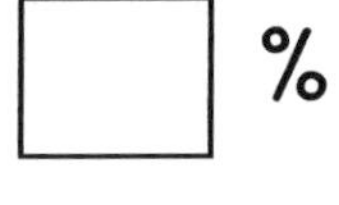 %

$$\rightarrow \frac{p}{100} = \frac{\square}{\square} \begin{matrix} \leftarrow \textit{PART} \\ \leftarrow \textit{WHOLE} \end{matrix}$$

Answer:

2. What percent of 40 is 25?  %

$$\frac{p}{\square} = \frac{\square}{\square}$$

Answer:

3. What percent of 28 is 12? □ %

$$\frac{\square}{\square} = \frac{\square}{\square}$$

Answer:

4. What percent of 79 is 68? 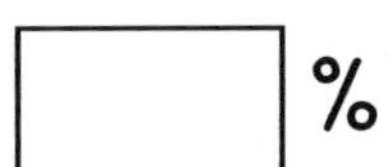 %

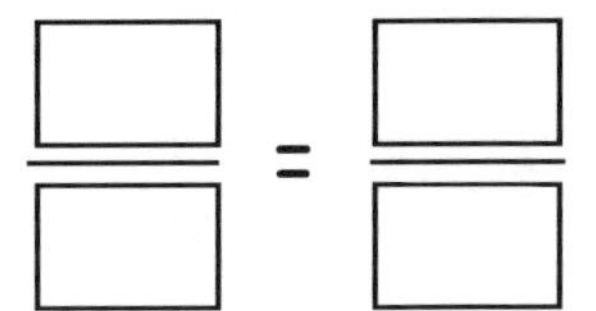

Answer:

5. What percent of 56 is 18?  %

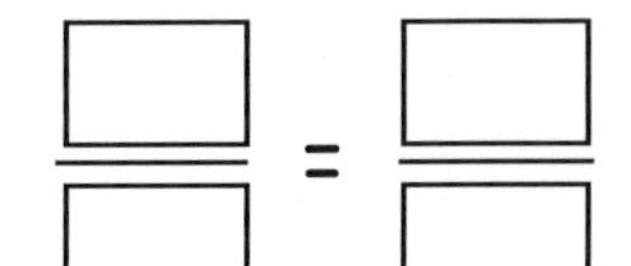

Find the % - Fraction, Decimal, Percent Method

Find the PERCENT, going from FRACTION to DECIMAL to PERCENT

Example

A. What percent of 28 is 7?

#1 Make a FRACTION out of the two numbers you have

As before, the number right after the word "of" is the denominator

$$\frac{7}{28}$$

#2 Convert to a DECIMAL by dividing . . .

TOP divided by the BOTTOM
(You may use a calculator on this part)

7 ÷ 28 = .25

#3 Convert to a PERCENT by using the DP Trick

Slide the decimal to the RIGHT two times

Stick on the percent sign! You're done!

25%

	FRACTION	DECIMAL	PERCENT
	Number after "of" is the denominator	Slide this decimal to the RIGHT twice	Round to nearest tenths place:

1. What percent of 45 is 9?

FRACTION: ☐ / ☐ DECIMAL: ☐ Answer: ☐ %

2. What percent of 82 is 56?

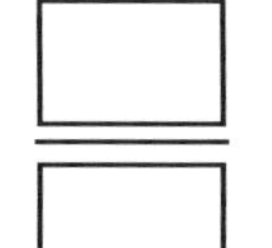

DECIMAL: ☐ Answer: ☐ %

Round this number to 3 decimal places

Find the % - Fraction, Decimal, Percent Method

Find the PERCENT, going from FRACTION to DECIMAL to PERCENT

	Number after "of" is the denominator FRACTION	Slide this decimal to the RIGHT twice DECIMAL	Round to nearest tenths place: PERCENT
1. What percent of 74 is 23?		Round this number to 3 decimal places	Answer: ___ %
2. What percent of 27 is 94?		Round this number to 3 decimal places	Answer: ___ %
3. What percent of 65 is 65?			Answer: ___ %
4. What percent of 40 is 80?			Answer: ___ %
5. What percent of 32 is 85?		Round this number to 3 decimal places	___ %
6. What percent of 98 is 82?		Round this number to 3 decimal places	Answer: ___ %

Find the Whole

Example

This is the proportion method:

PART!

WHOLE!

LOOK FOR THE
KEY WORD "OF"!

56 is 70% of what number ?

The value after this goes on the bottom of the fraction!

$$\frac{\text{PART}}{\text{WHOLE}} \quad \frac{70}{100} = \frac{56}{x} \quad \frac{\text{PART}}{\text{WHOLE}}$$

#1 Cross multiply $70x = 5600$

#2 Divide both sides to isolate the variable $\frac{70x}{70} = \frac{5600}{70}$

$x = 80$ 56 is 70% of 80

1. 32 is 80% of what number? ☐

$$\frac{\text{PART} \rightarrow}{\text{WHOLE} \rightarrow} \quad \frac{\square}{\square} = \frac{\square}{\square}$$

2. 14 is 20% of what number? ☐

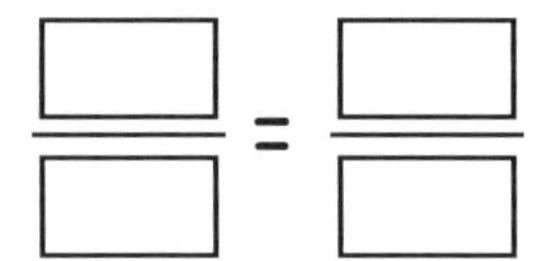

3. 12 is 8% of what number? ☐

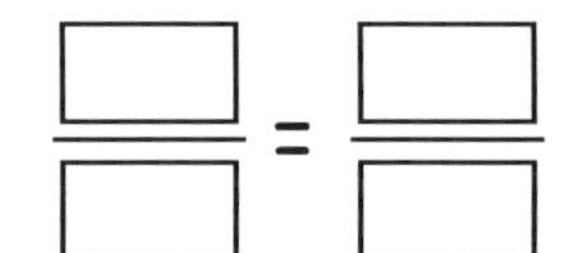

Find the Whole

Find the WHOLE, by setting up a proportion and solving for x

Example

A. 12 is 78% of what number?

$$\frac{78}{100} = \frac{12}{x}$$ ← PART / ← WHOLE

THE NUMBER AFTER "OF" GOES ON THE BOTTOM— HERE YOU PUT A LETTER!

#1 Set up a proportion

#2 Cross multiply $78x = 1200$

#3 Get the letter by itself by dividing both sides by the number in front of the letter!

$78x = 1200$

$\div 78$ $\div 78$

$x = 15.38$

#4 Round to the nearest tenth
Stick on the percent sign!

Answer: 15.4 %

1. 37 is 65% of what number?

#1 Set up a proportion

☐ / ☐ = ☐ / ☐ ← PART / ← WHOLE

#2 Cross multiply ☐

#3 Get the letter by itself by dividing both sides by the number in front of the letter!

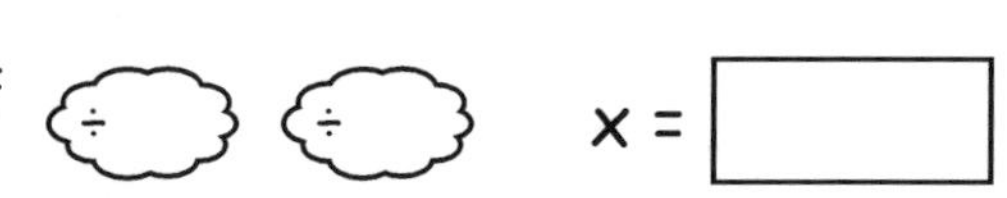

#4 Round to the nearest tenth
Stick on the percent sign!

Answer: ☐ %

Find the Whole

Find the WHOLE, by setting up a proportion and solving for x

1. 84 is 145% of what number?

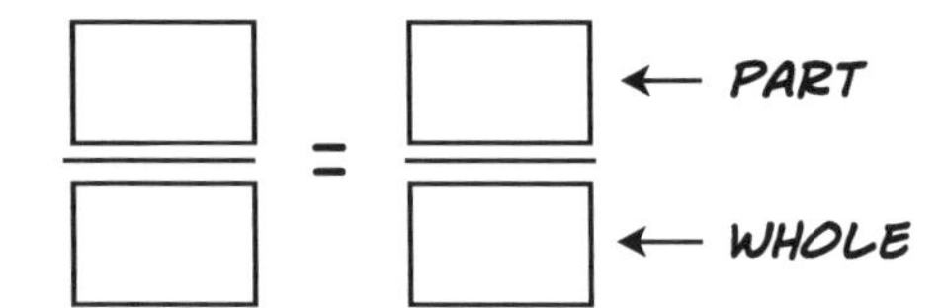

Answer:

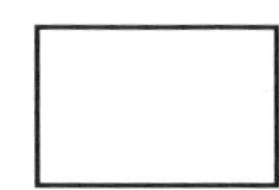

Round to the nearest tenth

2. 25 is 35% of what number?

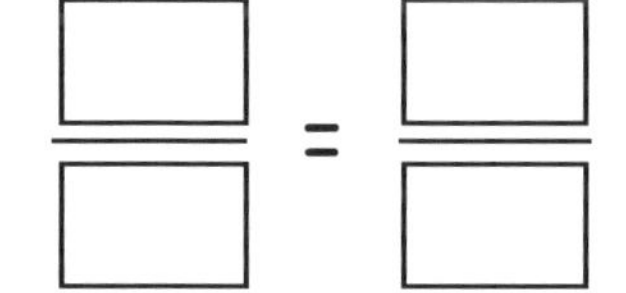

Answer:

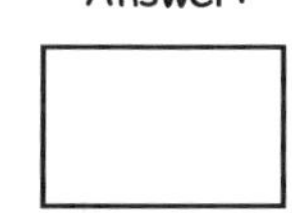

Round to the nearest tenth

3. 57 is 142% of what number?

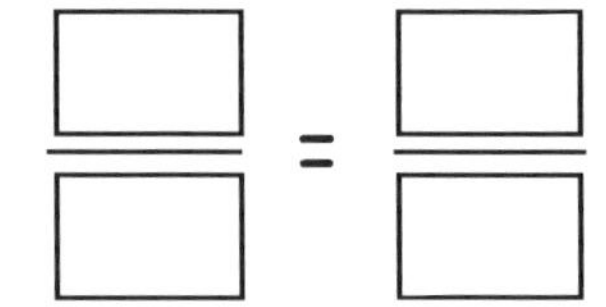

Answer:

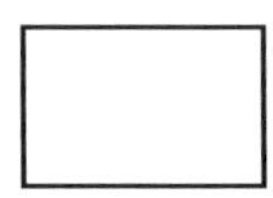

Round to the nearest tenth

Percents Test

1. 20% of 35 =

2. 75% of 32 =

3. $66\frac{2}{3}$ % of 12 =

4. Write 60% as a decimal

5. Write 7% as a decimal

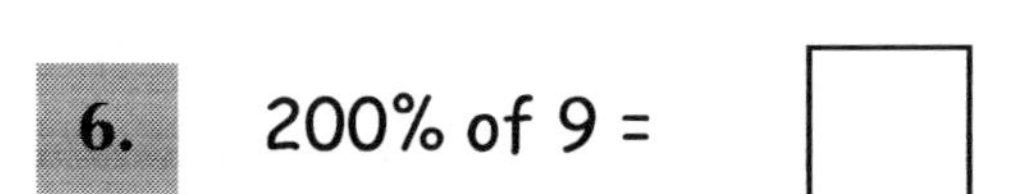

6. 200% of 9 =

7. What is 3% of 120?

8. Multiply by $\frac{1}{3}$ = ____ %

9. Calculate the cost of a $40 pair of shoes with a 10% discount

10. Calculate the cost of a $70 meal with a 15% tip

11. Sammy bought four shirts at $4.50 each and three pairs of pants at $10 each. What was the cost of all the clothes and a 10% tax?

12. Calculate the cost of a $20 binder with 6% tax

Calculating Percent of Change

Change Over Original

Example

A. Tommy shot 10 baskets at the basketball court on Monday. On Wednesday he shot 15 baskets. What was the percent increase?

#1 CHANGE / ORIGINAL → #2 CONVERT TO A DECIMAL → #3 CONVERT TO A PERCENT

YOU ALWAYS **SUBTRACT** TO MEASURE CHANGE (DIFFERENCE)!

$\frac{15 - 10}{10} = \frac{5}{10} =$ → .5 → 50%

MAKE SURE TO REPORT YOUR ANSWER HERE AS A 50% ***INCREASE!***

Answer: Wednesday's performance was a 50% increase.

Example

B. Bob used to read 50 pages a night. Now he's reading 40 pages a night. What is the percent of change?

$\frac{50 - 40}{50} = \frac{10}{50} = \frac{1}{5}$ → .2 → 20%

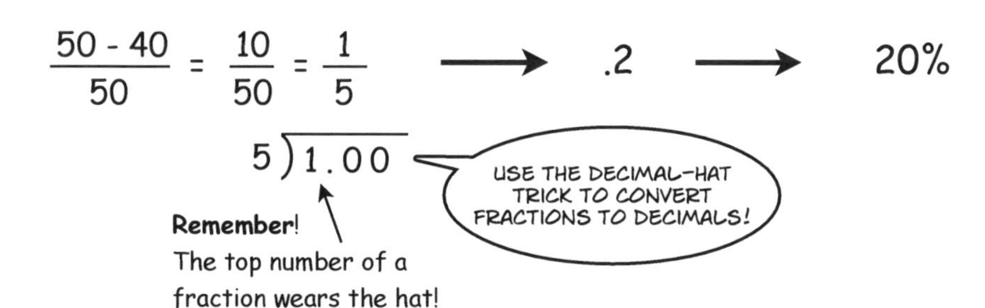

5) 1.00
USE THE DECIMAL-HAT TRICK TO CONVERT FRACTIONS TO DECIMALS!
Remember!
The top number of a fraction wears the hat!

Answer: Bob now reads 20% LESS than he did before.

IF THE AMOUNT IS DECREASING, BE SURE TO REPORT IT AS SUCH!
IN THIS CASE, USE "DECREASE" OR "LESS"!

Calculating Percent of Change

Change Over Original

1. On Wednesday the temperature was 80 degrees. On Thursday the temperature went up to 100 degrees. What % did it go up?

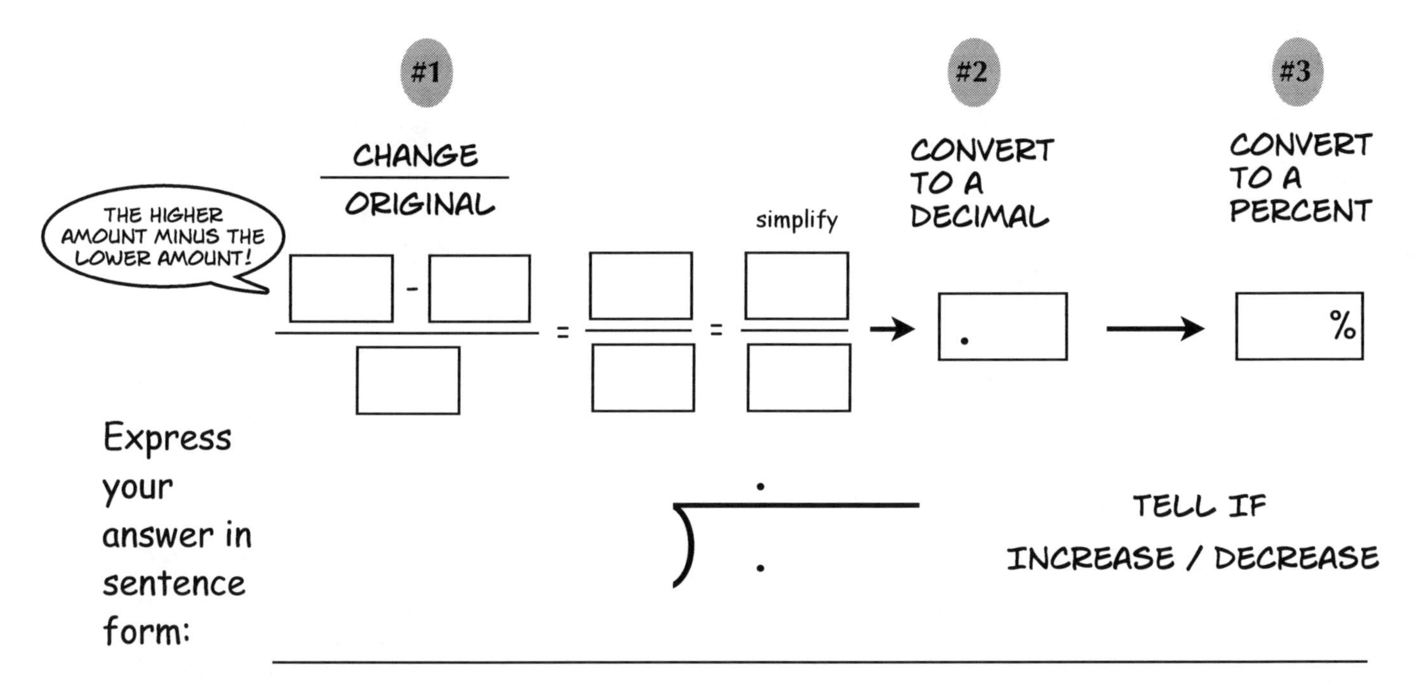

#1
CHANGE
ORIGINAL
#2
CONVERT TO A DECIMAL
#3
CONVERT TO A PERCENT
THE HIGHER AMOUNT MINUS THE LOWER AMOUNT!
simplify
%
Express your answer in sentence form:
TELL IF
INCREASE / DECREASE

2. Last year Brandon earned $15 an hour at his job. At his new job he makes only $9 an hour. What is the percent of change?

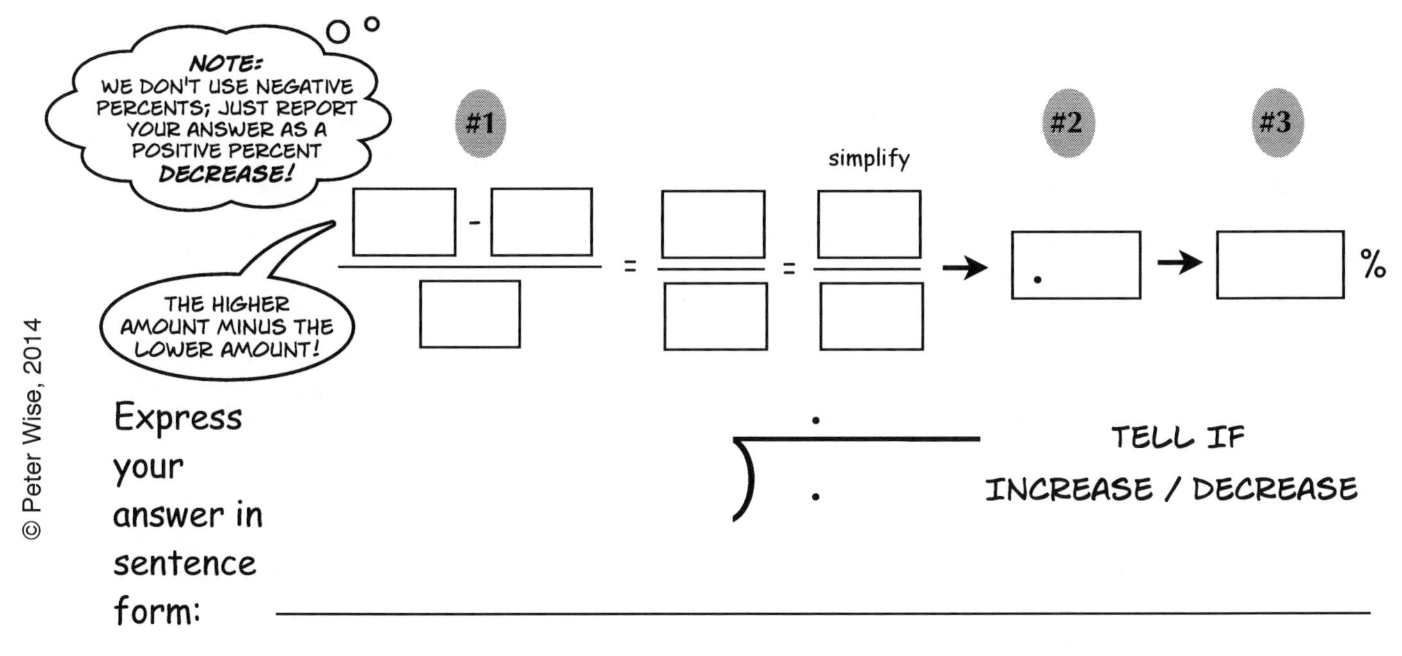

NOTE:
WE DON'T USE NEGATIVE PERCENTS; JUST REPORT YOUR ANSWER AS A POSITIVE PERCENT DECREASE!
#1
#2
#3
simplify
%
THE HIGHER AMOUNT MINUS THE LOWER AMOUNT!
Express your answer in sentence form:
TELL IF
INCREASE / DECREASE

Calculating Percent of Change

1. Suzie used to finish her chores in 40 minutes. This week she took 45 minutes to finish the chores. What is the percent of change?

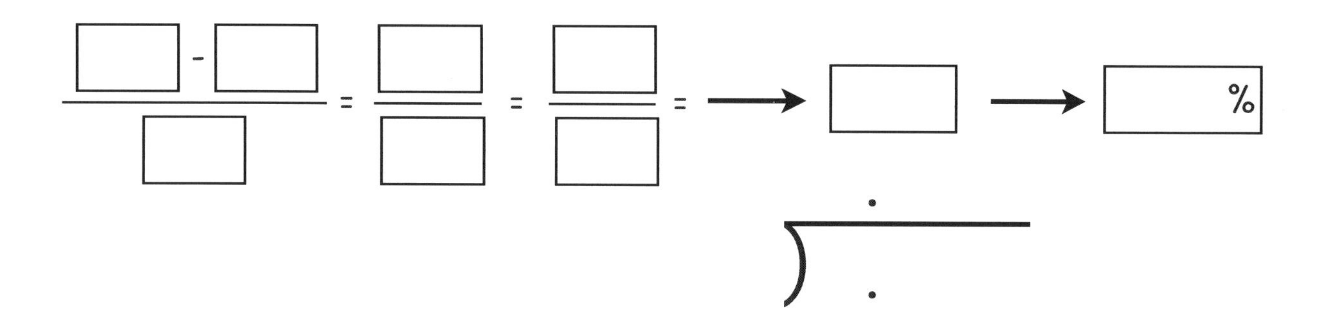

Express your answer in sentence form: ____________________

TELL IF
FASTER / SLOWER

2. Sandra's cleaning business made $160 on her first weekend. The second weekend she made $170. What percent increase or decrease did she have during the second week of operation?

Express your answer in sentence form: ____________________

TELL IF
INCREASE / DECREASE

Turning Fractions into Percents

Pretend that this division sign is a hat!

Put the division hat on the top number ("head") of the fraction → $\frac{3}{4}$

YOU MAY THINK THAT "F-D STANDS FOR "FIRE DEPARTMENT"! BUT IT REALLY STANDS FOR "FRACTION-DECIMAL"!

ONCE YOU KNOW WHICH NUMBERS GO ON THE INSIDE AND OUTSIDE OF THE DIVISION SIGN, DIVIDE NORMALLY!

Example

A. $\frac{1}{5}$ → $\frac{1}{5}$ → 5)1 5)1.000

SINCE 4 CAN'T GO INTO 3 EVENLY, ADD A DECIMAL POINT AND AS MANY ZEROS TO THE RIGHT AS YOU NEED!

IGNORE THE DECIMAL POINT, DIVIDE NORMALLY, AND THEN FLOAT THE DECIMAL TO THE TOP AT THE END!

```
   .2
5)1.00
 -1 0
    0
```

Since you're done you can skip the remaining zero(s)

So $\frac{1}{5}$ written as a decimal is **.2**

NOW DIVIDE AND FLOAT UP THE DECIMAL WHEN YOU'RE DONE!

1. $\frac{2}{5}$ ☐)☐ ☐☐ Answer: ☐

ADD AS MANY ZEROS AS YOU WANT OR NEED--TO THE RIGHT!

PUT THE NUMBER THAT WEARS THE HAT HERE!

ADD A DECIMAL POINT!

NOW USE THE "DR.DECIMAL-PEPPER PERCENT TRICK" TO CONVERT TO A PERCENT!

2. $\frac{3}{4}$) . Decimal: ☐ Percent: ☐ %

USE SCRATCH PAPER IF YOU NEED MORE SPACE!

3. $\frac{5}{8}$) . Decimal: ☐ Percent: ☐ %

Mixed Review Percent Problems

Solve the following percent problems; round to the nearest tenths place

CALCULATOR OKAY ON ALL BUT #1 AND #6

1. 87% of 50 =

2. What percent of 72 is 20?

3. 42% of what is 36?

4. What percent of 37 is 24?

5. 92 is what percent of 65?

6. .23% of 358 =

7. 73 is what percent of 16?

Solving Percents, using the Algebraic Method

Find the Percent

This is one of three basic types of percent problems. There are two ways of doing these:

- The proportion method (quicker, but easier to make a mistake with these)
- The algebraic method (longer, but some people find these easier to visualize)

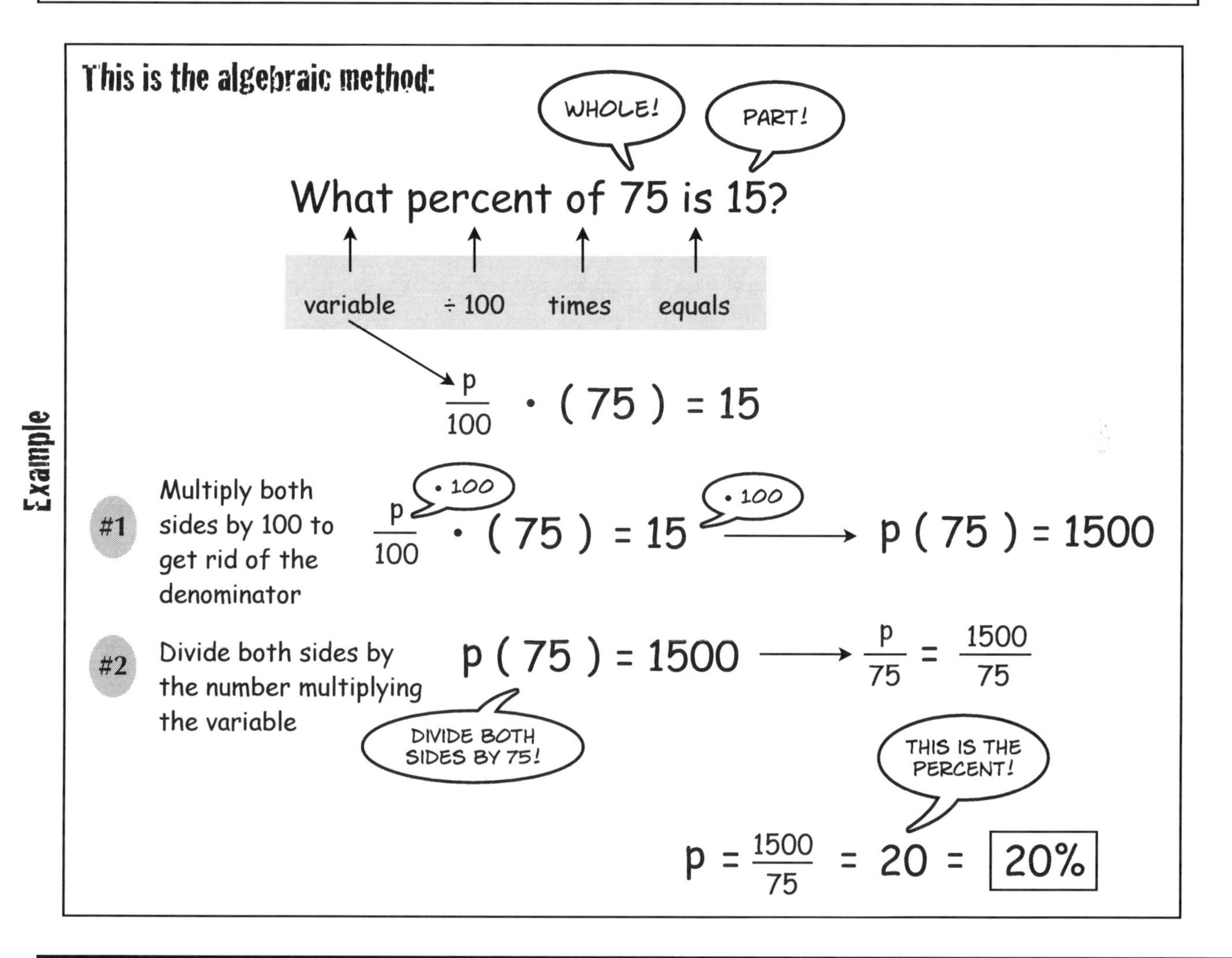

Solve the following percent problem

1. What percent of 50 is 4?

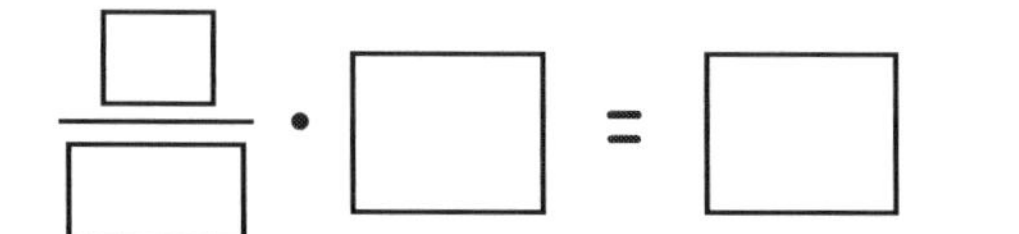

$= \square\ \%$ Answer:

#1 CANCEL FROM THE FRACTION AND THE 50!

#2 MULTIPLY BOTH SIDES BY THE NUMBER IN THE DENOMINATOR

Find the Percent

Calculate the following percent problems using the ALGEBRAIC method

1. What percent of 24 is 18? ☐ %

$$\frac{p}{100} \cdot 24 = 18$$

#1 MULTIPLY BOTH SIDES BY 100!

#2 THEN DIVIDE BOTH SIDES BY THE NUMBER MULTIPLYING THE VARIABLE!

$$\frac{24p}{24} = \frac{1800}{24}$$

$$p = 75$$

CALCULATOR OKAY ON THESE

Round to the nearest hundredth

2. What percent of 80 is 25? ☐ %

$$\frac{\square}{\square} \cdot \square = \square$$

MULTIPLY BOTH SIDES BY 100!

NEXT, DIVIDE BOTH SIDES BY THE NUMBER MULTIPLYING THE VARIABLE!

Round to the nearest hundredth

3. What percent of 90 is 15? ☐ %

$$\frac{\square}{\square} \cdot \square = \square$$

4. What percent of 40 is 35? ☐ %

$$\frac{\square}{\square} \cdot \square = \square$$

Find the Percent

Calculate the following percent problems using the ALGEBRAIC method

CALCULATOR OKAY ON THESE PROBLEMS

ROUND THE PERCENTS TO THE NEAREST TENTH!

1. What percent of 40 is 50? ☐ %

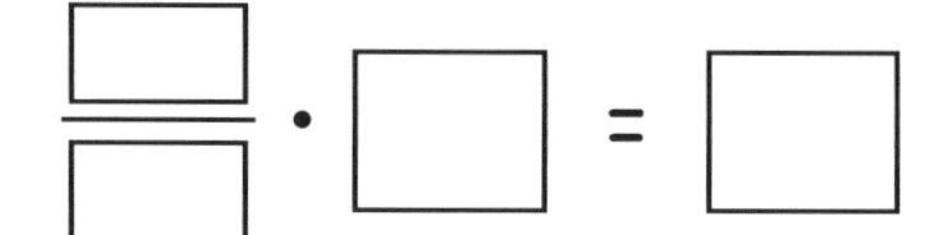

2. What percent of 48 is 34? ☐ %

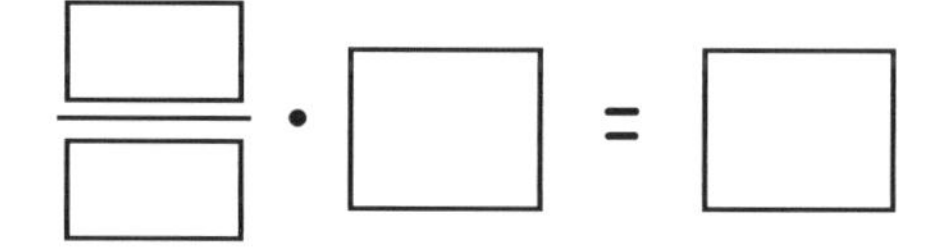

3. What percent of 75 is 20? ☐ %

4. What percent of 64 is 52? ☐ %

5. What percent of 84 is 37? ☐ %

Find the Whole

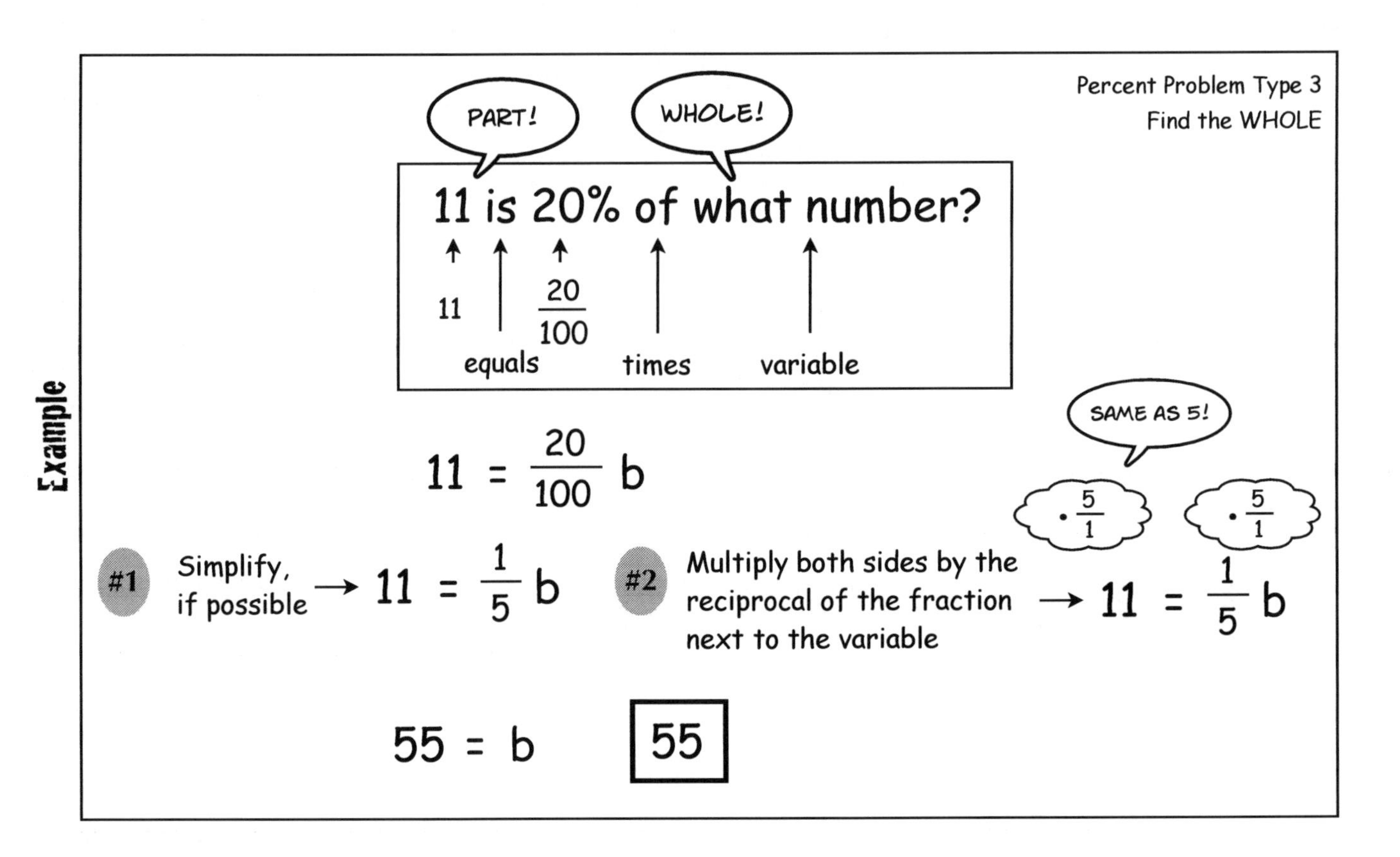

1. 6 is 25% of what number?

$\square = \frac{\square}{\square}\square$ $\quad$ $\square = \frac{\square}{\square}\square$ $\quad$ x = $\square$

Simplify and recopy

2. 12 is 75% of what number?

$\square = \frac{\square}{\square}\square$ $\quad$ $\square = \frac{\square}{\square}\square$ $\quad$ x = $\square$

Simplify and recopy

3. 9 is 15% of what number?

$\square = \frac{\square}{\square}\square$ $\quad$ $\square = \frac{\square}{\square}\square$ $\quad$ x = $\square$

Find the Whole

A. 14 is 64% of what number? **21.9**

$$\frac{64}{100} = \frac{16}{25}$$

Simplify and the percent first

$$\left(\frac{25}{16}\right) 14 = \frac{16}{25} \left(\frac{25}{16}\right) x$$

MULTIPLY BOTH SIDES BY THE RECIPROCAL TO CANCEL OUT 16/25!

$$\left(\begin{matrix}\cdot 25 \\ \div 16\end{matrix}\right) 14 = x$$

MULTIPLY 14 BY THE NUMERATOR, THEN DIVIDE BY THE DENOMINATOR!

$14 \cdot 25 = 350$

$350 \div 16 = 21.875$

21.875 rounds to 21.9

Calculate the following percent problems using the ALGEBRAIC method

1. 8 is 40% of what number?

THE SIMPLIFIED PERCENT FRACTION GOES HERE!

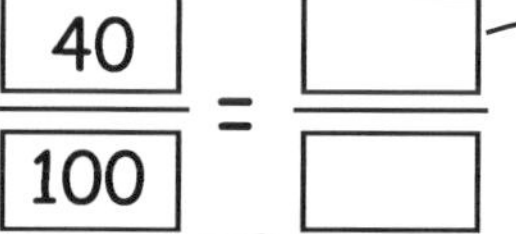

$$\frac{40}{100} = \frac{\square}{\square}$$

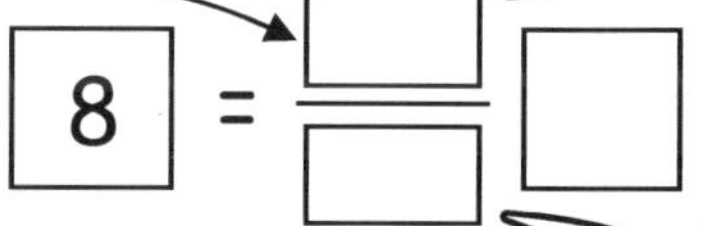

$$8 = \frac{\square}{\square} \square$$

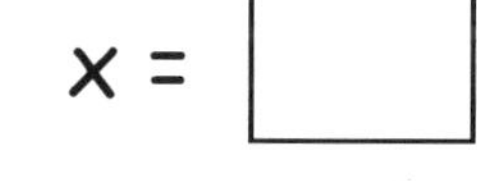

$x = \square$

SIMPLIFY THE PERCENT FRACTION AND PUT IT HERE!

ELIMINATE 2/5 BY MULTIPLYING BOTH SIDES BY THE RECIPROCAL (5/2)!

2. 27 is 45% of what number?

$$\frac{\square}{100} = \frac{\square}{\square} \qquad \square = \frac{\square}{\square} \square \qquad x = \square$$

Simplify and the percent first

3. 58 is 110% of what number?

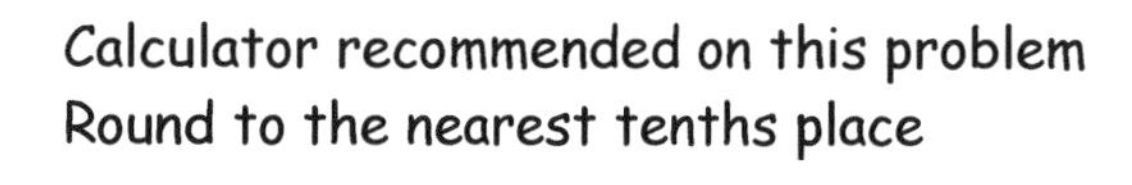

Calculator recommended on this problem
Round to the nearest tenths place

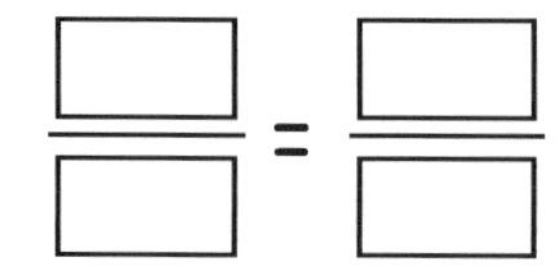

$$\frac{\square}{\square} = \frac{\square}{\square} \qquad \square = \frac{\square}{\square} \square \qquad x = \square$$

Find the Whole

Solve the following percent problems using the ALGEBRAIC method

CALCULATOR OKAY ON THESE

ROUND THE PERCENTS TO THE NEAREST TENTH!

1. 6 is 20% of what number?

$\frac{20}{100} = \frac{\square}{\square}$ $\quad \square = \frac{\square}{\square}\square$ $\quad x = \square$

Simplify and the percent first

2. 24 is 30% of what number?

$\frac{\square}{\square} = \frac{\square}{\square}$ $\quad \square = \frac{\square}{\square}\square$ $\quad x = \square$

Simplify and the percent first

3. 75 is 120% of what number?

$\frac{\square}{100} = \frac{\square}{\square}$ $\quad \square = \frac{\square}{\square}\square$ $\quad x = \square$

Simplify and the percent first

4. 72 is 48% of what number?

$x = \square$

5. 18 is 14% of what number?

$x = \square$

Algebraic Method Percents Quiz

Use the ALGEBRAIC METHOD to set up equations and solve

CALCULATOR OKAY ON THIS PAGE AFTER YOU SET UP THE EQUATIONS

ROUND THE PERCENTS TO THE NEAREST TENTH!

1. What percent of 70 is 20?

 %

2. 8 is 15% of what number? x = ☐

3. 74 is 168% of what number? x = ☐

4. What percent of 37 is 82?

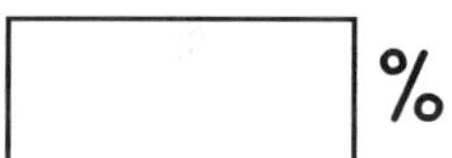 %

5. 26 is 73% of what number?

6. What percent of 95 is 71?

 %

Percents Comprehensive Test

1. 20% of 35 =

2. 80% of 40 =

3. $66.\overline{6}$% of 21 =

4. Write the number 11 as a percent:

5. Write 48% as a fraction in simplest form:

6. Write .6% as a decimal:

7. Write .095 as a percent:

8. Calculate the cost of a $70 meal with a 15% tip:

9. Calculate the cost of a $35 pair of shoes with a 10% discount:

10. Calculate the cost of a $60 jacket with 5% sales tax:

Percents Comprehensive Test

11. What is 27% of 64?

12. What percent of 24 is 5?

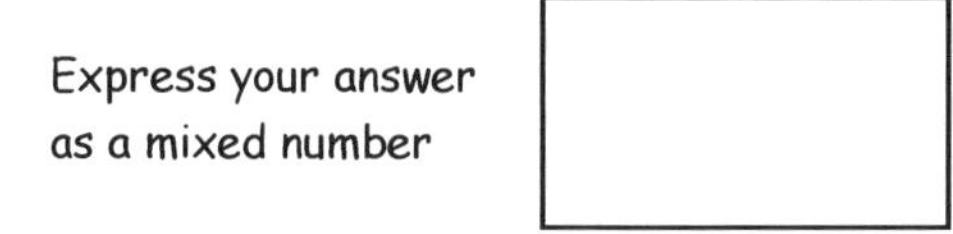

Express your answer as a mixed number

13. 38 is 70% of what number?

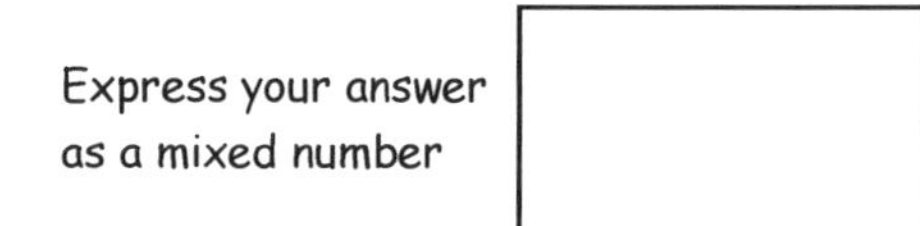

Express your answer as a mixed number

14. Tony's lawn care business made $150 on his first weekend. The second weekend he made $180. What percent increase or decrease did he have during the second week of operation?

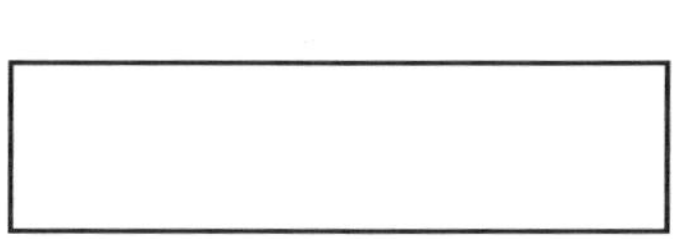

15. What is 2/7 as a percent?

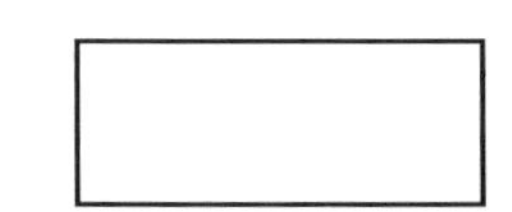

Round to the tenths place

Answer Key

for

MathWise Percents

What are Percents Anyway?

Percents are amounts out of 100

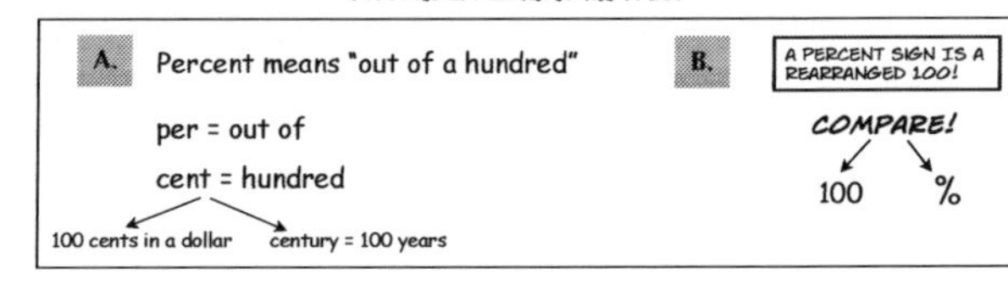

A. Percent means "out of a hundred"

per = out of

cent = hundred

100 cents in a dollar — century = 100 years

B. A PERCENT SIGN IS A REARRANGED 100!

COMPARE! 100 %

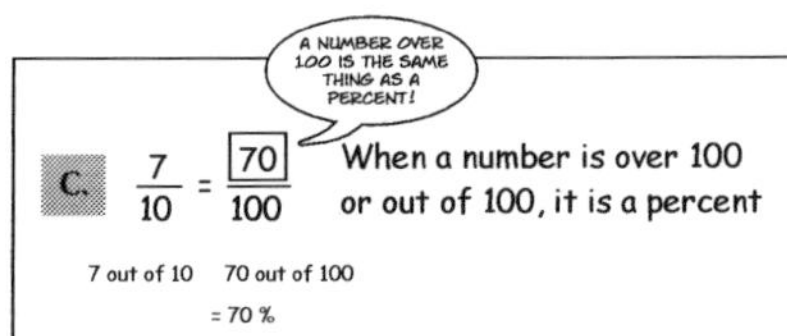

A NUMBER OVER 100 IS THE SAME THING AS A PERCENT!

C. $\frac{7}{10} = \frac{70}{100}$ When a number is over 100 or out of 100, it is a percent

7 out of 10 — 70 out of 100 = 70 %

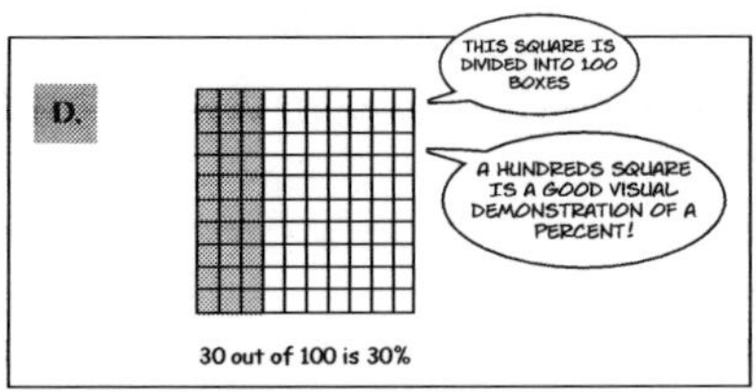

THIS SQUARE IS DIVIDED INTO 100 BOXES
A HUNDREDS SQUARE IS A GOOD VISUAL DEMONSTRATION OF A PERCENT!

D. 30 out of 100 is 30%

TWO DECIMAL SLIDES FOR THE TWO ZEROS IN THE PERCENT SIGN!

E. PERCENTS CAN BE WRITTEN AS DECIMALS! 35% is the same as .35

F. WHOLE NUMBERS CAN BE WRITTEN AS PERCENTS! 100% = 1; 400% = 4

Visualizing Percents

Example

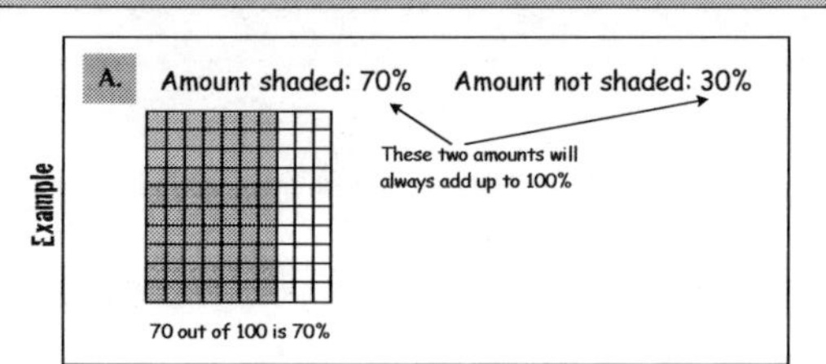

A. Amount shaded: 70% Amount not shaded: 30%

These two amounts will always add up to 100%

70 out of 100 is 70%

Tell or shade the correct amounts

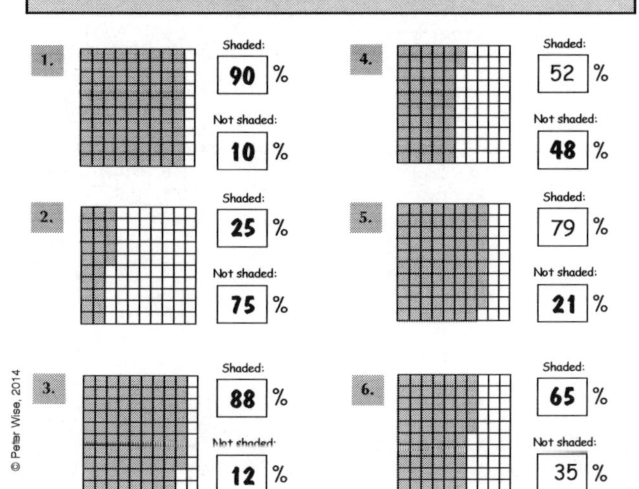

	Shaded	Not shaded
1.	90 %	10 %
2.	25 %	75 %
3.	88 %	12 %
4.	52 %	48 %
5.	79 %	21 %
6.	65 %	35 %

Determining Basic Percents

Divide into 100 to figure out the following percents

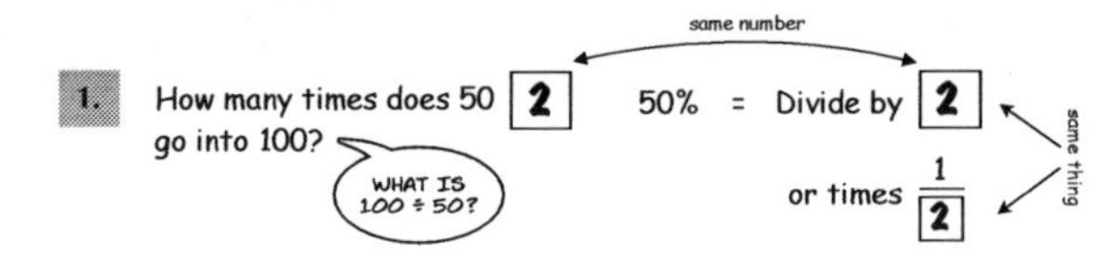

1. How many times does 50 go into 100? **2** — 50% = Divide by **2** or times $\frac{1}{2}$
2. How many times does 25 go into 100? **4** — 25% = Divide by **4** or times $\frac{1}{4}$
3. How many times does 20 go into 100? **5** — 20% = Divide by **5** or times $\frac{1}{5}$
4. How many times does 10 go into 100? **10** — 10% = Divide by **10** or times $\frac{1}{10}$
   or one decimal slide to the LEFT: 10% of 40 = 4.0 or 4
5. How many times does 5 go into 100? **20** — 5% = Divide by **20** or times $\frac{1}{20}$
6. How many times does 1 go into 100? **100** — 1% = Divide by **100** or times $\frac{1}{100}$
   or two decimal slides to the LEFT: 1% of 70 = .70 or .7

Basic Percent Values

Percents to Memorize

- **A.** 100% = Divide by 1 or times 1
- **B.** 50% = Divide by 2 or times $\frac{1}{2}$
- **C.** 25% = Divide by 4 or times $\frac{1}{4}$
- **D.** 20% = Divide by 5 or times $\frac{1}{5}$
- **E.** 10% = Divide by 10 or times $\frac{1}{10}$
- **F.** 1% = Divide by 100 or times $\frac{1}{100}$
- **G.** 5% = Divide by 20 or times $\frac{1}{20}$
- **H.** 15% = 10% + 5% (half of 10%)
- **I.** $33.\overline{3}\%$ = Divide by 3 or times $\frac{1}{3}$ — $33\frac{1}{3}\%$
- **J.** $66.\overline{6}\%$ = Divide by 3 and times 2 — $66\frac{2}{3}\%$ or times $\frac{2 \leftarrow \times}{3 \leftarrow \div}$
- **K.** 75% = Divide by 4 and times 3 or times $\frac{3 \leftarrow \times}{4 \leftarrow \div}$

Calculate the percentages of the following numbers

1. 100% of 45 = **45**
2. 50% of 60 = **30**
3. 25% of 12 = **3**
4. 20% of 35 = **7**
5. $33.\overline{3}\%$ of 18 = **6**
6. $66.\overline{6}\%$ of 24 = **16**
7. 10% of 90 = **9**
8. 1% of 80 = **.8**
9. 5% of 80 = **4**
10. 25% of 32 = **8**
11. 20% of 40 = **8**
12. 10% of 40 = **4**

Percent Practice

Calculate the percentages of the following numbers

1. 50% of 40 = **20**
2. 25% of 40 = **10** — 25% IS HALF OF 50%!
3. 75% of 40 = **30** — 75% IS JUST 50% + 25%. ...SO JUST ADD THE PREVIOUS TWO!
4. 50% of 36 = **18**
5. 25% of 36 = **9**
6. 75% of 36 = **27**
7. 20% of 55 = **11**
8. 10% of 60 = **6**
9. (+) 5% of 60 = **3** — 5% IS HALF OF 10%!
10. (=) 15% of 60 = **9** — 15% IS 10% + 5%!
11. 10% of 20 = **2**
12. 50% of 26 = **13**
13. 25% of 32 = **8**
14. 20% of 15 = **3**
15. 10% of 90 = **9**
16. 25% of 24 = **6**
17. 20% of 60 = **12**
18. 10% of 70 = **7**
19. 30% of 70 = **21** — 30% IS 3 TIMES WHAT 10% IS!
20. 80% of 70 = **56** — 80% IS 8 TIMES WHAT 10% IS!

PERCENT = DIVIDE BY 100!

THE TWO ZEROS ON THE NUMBERS MEAN "TIMES 100"!

NOTICE THAT YOU CAN CANCEL TWO PAIRS OF ZEROS—ONE ZERO ON THE PERCENT SIGN WITH EACH ZERO ON THE NUMBERS!

10% and 1% as Decimal Slides

Examples

- **A.** 10% of 37 = 3.7
- **B.** 1% of 37 = .37

REMEMBER THAT THERE IS AN INVISIBLE DECIMAL PLACE TO THE RIGHT OF THE ONE'S PLACE!

10% MEANS "DIVIDE BY 10!" THIS IS THE SAME AS SLIDING THE DECIMAL ONE TIME TO THE LEFT!

1% MEANS "DIVIDE BY 100!" THIS IS THE SAME AS SLIDING THE DECIMAL *TWO TIMES* TO THE *LEFT*!

Calculate the percentages of the following numbers

1. 10% of 58 = **5.8**
2. 1% of 63 = **.63**
3. 10% of 30 = **3**
4. 50% of 18 = **9**
5. 25% of 28 = **7**
6. 20% of 50 = **10**
7. 10% of 50 = **5**
8. 5% of 20 = **1**
9. 10% of 72 = **7.2**
10. 1% of 358 = **3.58**
11. 5% of 40 = **2**
12. 1% of 25 = **.25**
13. 10% of 70 = **7**
14. 1% of 70 = **.7**
15. 10% of 47 = **4.7**
16. 1% of 47 = **.47**
17. 50% of 14 = **7**
18. 20% of 15 = **3**
19. 10% of 248 = **24.8**
20. 5% of 100 = **5**

Percent Practice

Calculate the percentages of the following numbers

1. $33.\overline{3}\%$ of 60 = **20**
2. 1% of 500 = **5**
3. 1% of 365 = **3.65**
4. 50% of 28 = **14**
5. 20% of 45 = **9**
6. $33.\overline{3}\%$ of 15 = **5**
7. $66.\overline{6}\%$ of 15 = **10**
8. 10% of 80 = **8**
9. (+) 5% of 80 = **4**
10. (=) 15% of 80 = **12** — 15% IS JUST 10% + 5%!
11. 75% of 40 = **30**
12. 10% of 120 = **12**
13. 20% of 30 = **6**
14. 25% of 32 = **8**
15. $33.\overline{3}\%$ of 27 = **9**
16. 1% of 53 = **.53**
17. $66.\overline{6}\%$ of 60 = **40**
18. 20% of 25 = **5**
19. 1% of 30 = **.3**
20. 4% of 30 = **1.2** — 4% IS JUST 4 TIMES WHAT 1% IS!

Concept Quiz

1. A percent is an amount out of **a hundred**

2. Break down the word "percent": per = "**out of**"; cent = "**hundred**".

3. Shade in the correct amount and fill in the percentage not shaded:

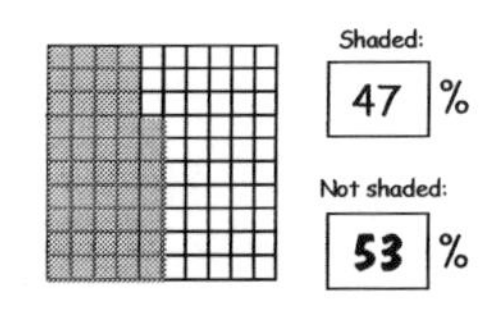

Shaded: **47** %

Not shaded: **53** %

4. If you can't remember what you divide by to get 25% of a number, what can you do to figure it out?
divide 100 by 25; 25% means divide by 4

5. What is the easy trick to figure out 10% of a number?
slide the decimal ONCE to the LEFT; or remove one zero

6. What is the easy trick to figure out 1% of a number?
slide the decimal TWICE to the LEFT; or remove 2 zeros

7. If you know 1% of a number, how could you easily calculate 2% of the same number?
double the amount you got by figuring out 1%

The DP Trick

DR. DECIMAL — 2 slides to the **LEFT**, if you have a PERCENT and want to make a DECIMAL

It takes 2 slides ⟷ ⟷ to get from one side to another!

PEPPER PERCENT — 2 slides to the **RIGHT**, if you have a DECIMAL and want to make a PERCENT

Use the DP Trick to change percents to decimals and the other way around

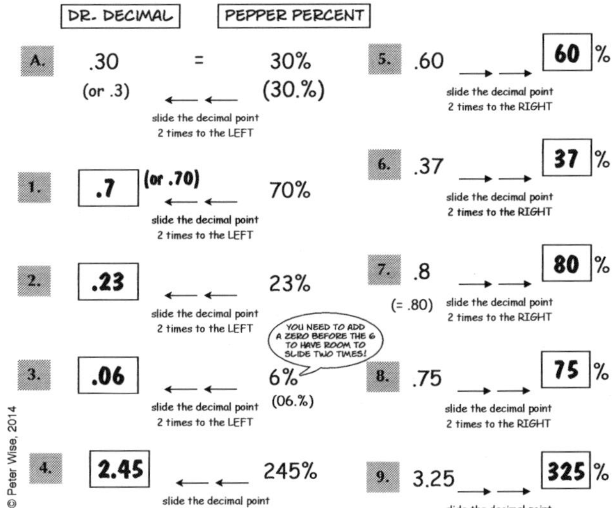

	DR. DECIMAL		PEPPER PERCENT
A.	.30 (or .3)	= ← ← slide the decimal point 2 times to the LEFT	30% (30.%)
1.	**.7** (or .70)	← ← slide the decimal point 2 times to the LEFT	70%
2.	**.23**	← ← slide the decimal point 2 times to the LEFT	23%
3.	**.06**	← ← slide the decimal point 2 times to the LEFT	6% (06.%)
4.	**2.45**	← ← slide the decimal point 2 times to the LEFT	245%

	DR. DECIMAL		PEPPER PERCENT
5.	.60	→ → slide the decimal point 2 times to the RIGHT	**60** %
6.	.37	→ → slide the decimal point 2 times to the RIGHT	**37** %
7.	.8 (= .80)	→ → slide the decimal point 2 times to the RIGHT	**80** %
8.	.75	→ → slide the decimal point 2 times to the RIGHT	**75** %
9.	3.25	→ → slide the decimal point 2 times to the RIGHT	**325** %

The DP Trick

Just slide two times to go from decimals to percents (or percents to decimals)

Use the DP Trick to change percents to decimals or the other way around

	DR. DECIMAL		PEPPER PERCENT
1.	.26	=	**26** %
2.	**.73**	=	73%
3.	**.8**	=	80%
4.	.4	=	**40** %
5.	.05	=	**5** %
6.	9	=	**900** %
7.	**2.47**	=	247%
8.	**.06**	=	6%
9.	.045	=	**4.5** %
10.	5.7	=	**570** %
11.	**.006**	=	.6%
12.	**3.21**	=	321%
13.	.002	=	**.2** %
14.	36	=	**3600** %
15.	**.12**	=	12%
16.	**.273**	=	27.3%
17.	3.14	=	**314** %
18.	**5**	=	500%

Percents as Division & Multiplication

Give the percent equivalents

1. Divide by 2 = **50** %
2. Divide by 4 = **25** %
3. Divide by 5 = **20** %
4. Divide by 10 = **10** %
5. Divide by 20 = **5** %
6. Divide by 100 = **1** %
7. Divide by 1 = **100** %
8. ÷ by 3 or $\cdot \frac{1}{3}$ = **$33.\overline{3}$** %
9. Times $\frac{2}{3}$ = **$66.\overline{6}$** %
10. Times $\frac{3}{4}$ = **75** %

VIEWED A SLIGHTLY DIFFERENT WAY...

11. 18 ÷ 2 is the same as **50** % of 18
12. 40 ÷ 10 is the same as **10** % of 40
13. 35 ÷ 5 is the same as **20** % of 35
14. 12 ÷ 3 is the same as **$33.\overline{3}$** % of 12
15. 32 ÷ 4 is the same as **25** % of 32
16. 24 ÷ 3 · 2 is the same as **$66.\overline{6}$** % of 24

Percent Review

1. 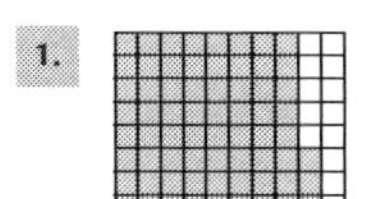
Shaded: **85** %
Not shaded: **15** %

2. How many times does 20 go into 100? **5**
20% = Divide by **5**
or times $\frac{1}{\boxed{5}}$

3. Divide by 4 = **25** %

DR. DECIMAL		PEPPER PERCENT
4. .30	=	**30** %
5. **.72**	=	72%
6. .09	=	**9** %
7. **.08**	=	8%
8. 2.5	=	**250** %

9. 25% of 40 = **10**

10. Times $\frac{1}{3}$ = **$33.\overline{3}$** %

11. Times $\frac{2}{3}$ = **$66.\overline{6}$** %

12. Times $\frac{3}{4}$ = **75** %

13. 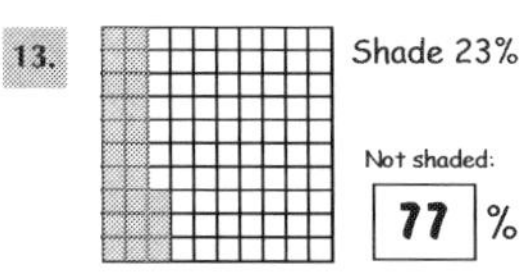
Shade 23%
Not shaded: **77** %

14. Divide by 10 = **10** %

15. Divide by 1 = **100** %

16. 70% = 10% × **7**

17. How does 5% differ from 10%?
5% is half of 10%

Percent Practice

Calculate the percentages of the following numbers.

1. 50% of 70 = **35**
2. 100% of 8 = **8**
3. 150% of 8 = **12**
4. 25% of 32 = **8**
5. 25% of 12 = **3**
6. 125% of 12 = **15**
7. 10% of 90 = **9**
8. 20% of 90 = **18**
9. 30% of 90 = **27**
10. 20% of 30 = **6**
11. 20% of 45 = **9**
12. 25% of 44 = **11**
13. 10% of 46 = **4.6**
14. 1% of 46 = **.46**
15. 10% of 60 = **6**
16. 100% of 60 = **60**
17. 110% of 60 = **66**
18. 50% of 16 = **8**
19. 25% of 16 = **4**
20. 1% of 365 = **3.65**

NOW TRY THESE PROBLEMS!

21. 10% of 30 = **3**
22. 70% of 30 = **21**
23. What is the price of a $40 meal plus a 15% tip? **$46**

Concept Quiz

1. Percents and decimals are just different ways of writing the same number: (True) / False (circle one)

2. How many slides does it take to go from a decimal to a percent (or the other way around)?
Two

3. Why does it take this number of slides? **Because % means out of 100 or because % means divided by 100; a % sign has 2 zeros**

4. Which way do you slide if you have a percent and want to write a decimal?
Slide to the (LEFT) / Slide to the RIGHT (circle one)

5. Divide by 5 is the same as WHAT PERCENT of a number? **20%**

6. How can you figure this out if you need to do so?
Divide 5 into 100

7. If you know 10% of a number, how can you easily calculate 30% of the same number?
Multiply 10% by 3

8. How does 5% differ from 10%?
5% is half of 10% or 10% is double 5%

Percents in Terms of Other Percents

Calculate, using basic operations with percents

1. 30% = 10% × **3**
2. 70% = 10% × **7**
3. 80% = 20% × **4**
4. 50% = 25% × **2**
5. 50% = 10% × **5**
6. 90% = 10% × **9**
7. 40% = 20% × **2**
8. 25% = 50% ÷ **2**
9. 5% = 50% ÷ **10**
10. 5% = 10% ÷ **2**
11. 100% = 20% × **5**
12. 20% = 10% × **2**
13. 60% = 10% × **6**
14. 60% = 20% × **3**
15. 15% = 10% + **5** %
16. 25% = 20% + **5** %
17. 15% = 20% - **5** %
18. 90% = 100% - **10** %
19. 80% = 100% - **20** %
20. 30% = 20% + **10** %

PUT THIS SKILL TO GOOD USE!

21. 35% of 50 =
35% is the sum of these percents
20% of 50 = **10**
10% of 50 = **5**
\+ 5% of 50 = **2.5**
35% **17.5**

Percent Review

1. Shaded: **38** % — Not shaded: **62** %

2. How many times does 5 go into 100? **20**

5% = Divide by **20**

or times $\frac{1}{20}$

3. Divide by 100 = **1** %

	DR. DECIMAL		PEPPER PERCENT
4.	.02	=	**2** %
5.	**1.25**	=	125%
6.	.57	=	**57** %
7.	**.08**	=	8%
8.	.006	=	**.6** %

9. 5% of 40 = **2**

10. Times $\frac{3}{4}$ = **75** %

11. Times $\frac{1}{3}$ = **$33.\overline{3}$** %

12. Times $\frac{2}{3}$ = **$66.\overline{6}$** %

13. Shade 83% — Unshaded: **17** %

14. 40% = 10% × **4**

15. 60% = 20% × **3**

16. $66.\overline{6}$% = $33.\overline{3}$% × **2**

17. What is one way you could calculate 45%?

Possible answers: 50% - 5% or 20% + 20% + 5%

Using 10% as a Reference

Use 10% as a reference to figure out other percent problems

Example

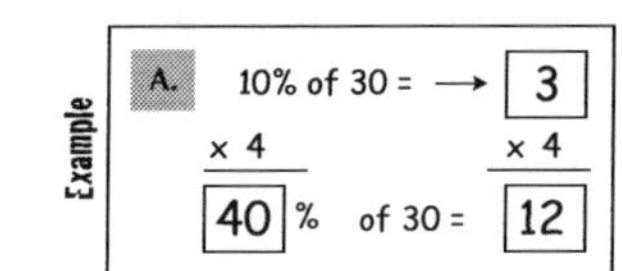

1. 10% of 70 = → **7**; × 3, × 3; **30** % of 70 = **21**

2. 10% of 60 = → **6**; × 9, × 9; **90** % of 60 = **54**

3. 10% of 50 = → **5**; × 7, × 7; **70** % of 50 = **35**

4. 10% of 120 = → **12**; × 8, × 8; **80** % of 120 = **96**

5. 10% of 20 = → **2**; × 6, × 6; **60** % of 20 = **12**

6. 10% of 40 = → **4**; × 7, × 7; **70** % of 40 = **28**

7. 10% of 80 = → **8**; × 6, × 6; **60** % of 80 = **48**

8. 10% of 50 = → **5**; ÷ 2, ÷ 2; **5** % of 50 = **2.5**

9. 10% of 30 = → **3**; ÷ 2, ÷ 2; **5** % of 30 = **1.5**

Percent Practice

Calculate the percentages of the following numbers

1. 10% of 45 = **4.5**

2. 1% of 45 = **.45**

3. $33\frac{1}{3}$% of 27 = **9**

4. $66\frac{2}{3}$% of 27 = **18**

5. $66.\overline{6}$% of 18 = **12**

6. $33.\overline{3}$% of 18 = **6**

7. 75% of 44 = **33**

8. 10% of 40 = **4**

+

9. 5% of 40 = **2**

=

10. 15% of 40 = **6**

15% IS JUST 10% + 5%!

11. 20% of 50 = **10**

12. $66\frac{2}{3}$% of 24 = **16**

13. 20% of 35 = **7**

14. 75% of 36 = **27**

15. $33.\overline{3}$% of 60 = **20**

16. 300% of 7 = **21**

17. $66.\overline{6}$% of 90 = **60**

18. 1% of 500 = **5**

19. 2% of 500 = **10**

20. 6% of 500 = **30**

COMPARE THESE TO 1%!

12.5% and Other Fraction Percents

Example

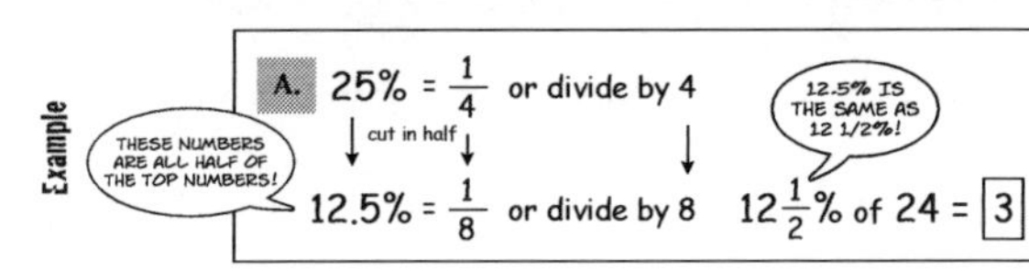

A. 25% = 1/4 or divide by 4
cut in half
12.5% = 1/8 or divide by 8
12 1/2% of 24 = 3
THESE NUMBERS ARE ALL HALF OF THE TOP NUMBERS!
12.5% IS THE SAME AS 12 1/2%!

Calculate the percentages of the following numbers

1. 12.5% of 40 = **5**

2. $33\frac{1}{3}$% of 60 = **20**

3. 75% of 28 = **21**

4. 12.5% of 32 = **4**

5. $66\frac{2}{3}$% of 33 = **22**

6. $33\frac{1}{3}$% of 21 = **7**

7. $12\frac{1}{2}$% of 72 = **9**

8. 700% of 6 = **42**

9. $66\frac{2}{3}$% of 30 = **20**

10. 12.5% of 64 = **8**

REVIEW PROBLEMS

11. Shade 49%

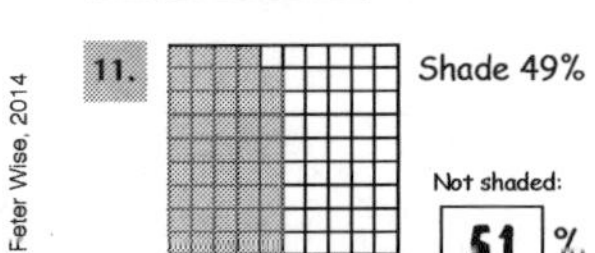

Not shaded: **51** %

12. 90% = 10% × **9**

13. Divide by 4 = **25** %

14. Divide by 8 = **12.5** %

Percent Practice

Calculate the percentages of the following numbers

1. 20% of 40 = **8**
2. 100% of 60 = **60**
3. $12\frac{1}{2}$% of 16 = **2**
4. $33\frac{1}{3}$% of 30 = **10**
5. 10% of 45 = **4.5**
6. 1% of 16 = **.16**
7. $66.\overline{6}$% of 18 = **12**
8. 300% of 20 = **60**
9. $12\frac{1}{2}$% of 24 = **3**
10. 50% of 48 = **24**
11. $66.\overline{6}$% of 21 = **14**
12. 25% of 80 = **20**
13. 5% of 80 = **4**
14. 10% of 6 = **.6**
15. 30% of 6 = **1.8**
16. 75% of 48 = **36**
17. 20% of 50 = **10**
18. 1% of 2000 = **20**
19. 3% of 2000 = **60**
20. 6% of 2000 = **120**

ANOTHER WAY TO CALCULATE THIS ONE IS TO REMEMBER THAT 5% IS FIVE TIMES SMALLER THAN 25%!

JUST 3 TIMES WHAT 10% OF 6 IS!

COMPARE THESE LAST TWO TO 1%!

100% and Above

Example

A. 1.00 = 100% Two zeros on the percent sign = Two slides on the decimal

In other words 100% is the same as the number 1

Write these amounts for percents that are greater than 1

1. 200% = **2**
2. 300% = **3**
3. 700% = **7**
4. 1000% = **10**
5. 800% = **8**
6. 12,000% = **120**
7. 400% = **4**
8. 600% = **6**
9. 900% = **9**

A LITTLE HARDER---

10. 200% of 9 = **18**
11. 300% of 7 = **21**
12. 500% of 9 = **45**
13. 800% of 2 = **16**
13. 700% of 4 = **28**
14. 600% of 3 = **18**
15. 100% of 9 = **9**
16. 1200% of 3 = **36**
17. 1100% of 6 = **66**

REMEMBER THAT "OF" MEANS "TIMES"

Percents by Making Denominators a Hundred

WHEN THE BOTTOM NUMBER IS 100, THE TOP NUMBER TELLS THE PERCENT!

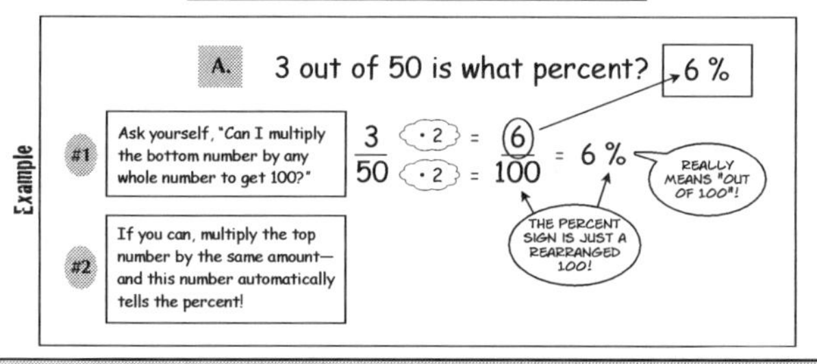

Example
A.
3 out of 50 is what percent?
6 %
#1
Ask yourself, "Can I multiply the bottom number by any whole number to get 100?"
#2
If you can, multiply the top number by the same amount—and this number automatically tells the percent!
3 · 2 = 6
50 · 2 = 100
= 6 %
REALLY MEANS "OUT OF 100"!
THE PERCENT SIGN IS JUST A REARRANGED 100!

Figure out the percent by making the bottom number equal to 100

1. 7 out of 10 is what percent?
   $\frac{7 \cdot 10}{10 \cdot 10} = \frac{70}{100}$ **70** %

2. What percent is 4 out of 5?
   $\frac{4 \cdot 20}{5 \cdot 20} = \frac{80}{100}$ **80** %

3. 15 out of 20 players on the team got base hits. What percent was that?
   $\frac{15 \cdot 5}{20 \cdot 5} = \frac{75}{100}$ **75** %

4. Twenty-five students are in Mr. Smith's class. Three of them were absent yesterday. What percent was that?
   $\frac{3 \cdot 4}{25 \cdot 4} = \frac{12}{100}$ **75** %

5. Five friends went out to eat. Three ordered cheeseburgers. What percent was this?
   $\frac{3 \cdot 20}{5 \cdot 20} = \frac{60}{100}$ **60** %

6. 14 out of 200 animals at the zoo bite the zookeepers regularly. What percent do this?
   $\frac{14 \div 2}{200 \div 2} = \frac{28}{100}$ **28** %

If the Bottom Number is 100

---THE TOP NUMBER TELLS THE PERCENT!

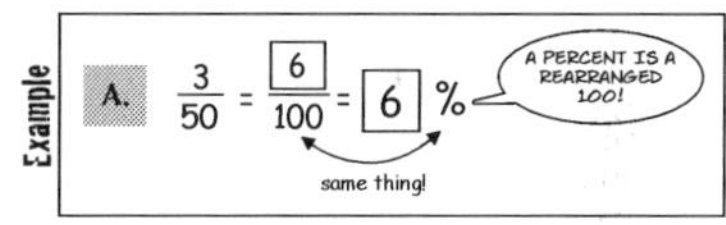

Calculate the percentages of the following numbers

1. $\frac{10}{25} = \frac{40}{100} =$ **40** % (same thing!)
2. $\frac{3}{20} = \frac{15}{100} =$ **15** %
3. $\frac{1}{4} = \frac{25}{100} =$ **25** %
4. $\frac{2}{4} = \frac{50}{100} =$ **50** %
5. $\frac{3}{4} = \frac{75}{100} =$ **75** %
6. $\frac{1}{5} = \frac{20}{100} =$ **20** %
7. $\frac{3}{5} = \frac{60}{100} =$ **60** %
8. $\frac{2}{10} = \frac{20}{100} =$ **20** %
9. 40 out of 50
   $\frac{20}{50}$ (out of) $= \frac{80}{100} =$ **80** %
10. 2 out of 5
   $\frac{2}{5}$ (out of) $= \frac{40}{100} =$ **40** %
11. 7 out of 25
   $\frac{7}{25}$ (out of) $= \frac{28}{100} =$ **28** %

Percents as Fractions

Example
A. 75% = 75/100
WHEN THE BOTTOM NUMBER'S A HUNDRED, THE TOP NUMBER TELLS THE PERCENT!
75 ÷25 / 100 ÷25 = 3/4
same thing, really
ALWAYS TRY TO SIMPLIFY YOUR ANSWER!

Convert the following percents to fractions

1. $20\% = \frac{20}{100} = \frac{\div 20}{\div 20} \frac{1}{5}$
2. $30\% = \frac{30}{100} = \frac{\div 10}{\div 10} \frac{3}{10}$
3. $50\% = \frac{50}{100} = \frac{\div 50}{\div 50} \frac{1}{2}$
4. $25\% = \frac{25}{100} = \frac{\div 25}{\div 25} \frac{1}{4}$
5. $10\% = \frac{10}{100} = \frac{\div 10}{\div 10} \frac{1}{10}$
6. $15\% = \frac{15}{100} = \frac{\div 5}{\div 5} \frac{3}{20}$
7. $35\% = \frac{35}{100} = \frac{\div 5}{\div 5} \frac{7}{20}$
8. $60\% = \frac{60}{100} = \frac{\div 20}{\div 20} \frac{3}{5}$
9. $40\% = \frac{40}{100} = \frac{\div 20}{\div 20} \frac{2}{5}$
10. $5\% = \frac{5}{100} = \frac{\div 5}{\div 5} \frac{1}{20}$
11. $120\% = \frac{120}{100} = 1\frac{1}{5}$ (write in simplest form)
12. $250\% = \frac{250}{100} = 2\frac{1}{2}$ (write in simplest form)

Percent Review

1. 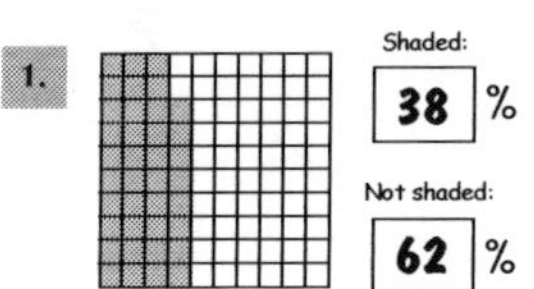
Shaded: 38 %
Not shaded: 62 %

2. 12.5% of 48 = 6

3. Divide by 25 = 4 %

4. $\frac{7}{20} = \frac{35}{100} = 35\%$

5. $60\% = \frac{60}{100} = \frac{3}{5}$

DR. DECIMAL		PEPPER PERCENT
6. .02	=	2 %
7. 1.25	=	125%
8. .57	=	57 %
9. .08	=	8%
10. .006	=	.6 %

11. 15 out of 50
$\frac{15}{50} = \frac{30}{100} = 30\%$

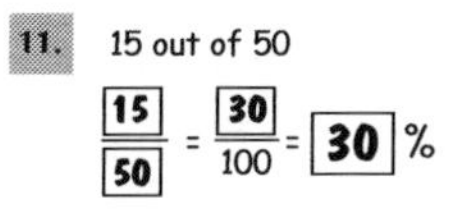

12. Shade 56%
Not shaded: ____ %

13. What percent is 2 out of 5?
$\frac{2 \cdot 20}{5 \cdot 20} = \frac{40}{100}$ = 40 %

14. 3 out of 25
$\frac{3}{25} = \frac{12}{100} = 12\%$

15. 75% = 25% × 3

16. 60% = 50% + 10 %

17. What is one way you could calculate 90%?

10% times 9
100% - 10%, etc.

Working with Percents

Calculate the percentages of the following numbers

1. 50% of 12 = 6
2. 150% of 12 = 18
3. 10% of 50 = 5
4. 5% of 50 = 2.5
5. 15% of 50 = 7.5
6. 20% of 35 = 7
7. 75% of 28 = 21
8. $33\frac{1}{3}\%$ of 60 = 20
9. $66\frac{2}{3}\%$ of 12 = 8
10. 25% of 36 = 9

Express the division problems as percents

11. Divide by 5 = 20 %
12. Divide by 10 = 10 %
13. Divide by 100 = 1 %
14. Divide by 2 = 50 %
15. Divide by 3 = $33\frac{1}{3}$ %
16. Divide by 3 and times by 2 = $66\frac{2}{3}$ %
17. Divide by 4 = 25 %
18. Divide by 4 and times by 3 = 75 %

Percent Practice

Solve the following percent problems

1. 50% of 16 = 8
2. 25% of 16 = 4
3. 75% of 16 = 12
4. 10% of 23 = 2.3
5. 10% of 10 = 1
6. 20% of 10 = 2
7. 10% of 120 = 12

8. 5% of 120 = 6

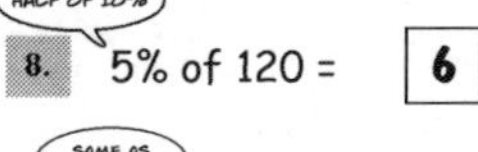

9. 15% of 120 = 18
10. 20% of 45 = 9
11. 50% is ÷ by 2
12. 25% is ÷ by 4
13. 20% is ÷ by 5
14. 10% is ÷ by 10
15. 5% is ÷ by 20
16. 33.3% is ÷ by 3
17. 66.6% is ÷ by and · by
Bottom of a fraction = divide
Top of a fraction = multiply
3 2
18. 75% is ÷ by and · by
4 3
19. 100% is ÷ by 1
20. 300% is · by 3

DP Trick Practice

Just slide two times to go from decimals to percents (or percents to decimals)

Use the DP Trick to change percents to decimals or the other way around

1. Write 8% as a decimal **.08**
2. Write 23% as a decimal **.23**
3. Write .3 as a percent **30** %
4. Write .42 as a percent **42** %
5. Write 20% as a decimal **.2**
6. Write 200% as a decimal **2**
7. Write .9 as a percent **90** %
8. Write 9 as a percent **900** %
9. Write .4% as a decimal **.004**
10. Write 8.2% as a decimal **.082**
11. Tommy tried to impress people with his age by telling people that his age was 1600%. How old is he really? **16**
12. Sally tried to confuse her friends by saying that she paid 75% of a dollar for a candy bar. How much was it really? **$0.75**
13. What is 15% as a decimal? **.15**
14. At a certain electronics store the prices increased by a factor of .25. What is that factor as a percent? **25%**

15% Restaurant Tips

Example

A. Calculate the cost of a $20 meal with a 15% tip

Cost of meal: $20

15% tip: + 10% of $20: $2 (cut this in half to get 5%)
5% of $20: $1 (add the 10% and the 5% to get 15%)
15% $3

Cost of meal plus the tip: $23

Calculate the cost of the meal PLUS A 15% TIP

1. Calculate the cost of a $60 meal with a 15% tip

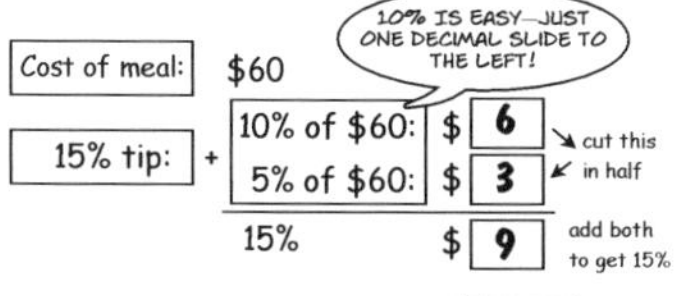

10% IS EASY—JUST ONE DECIMAL SLIDE TO THE LEFT!

Cost of meal: $60

15% tip: + 10% of $60: $ **6** (cut this in half)
5% of $60: $ **3**
15% $ **9** (add both to get 15%)

Cost of meal plus the 15%tip: **$69**

2. Calculate the cost of a $50 meal with a 15% tip

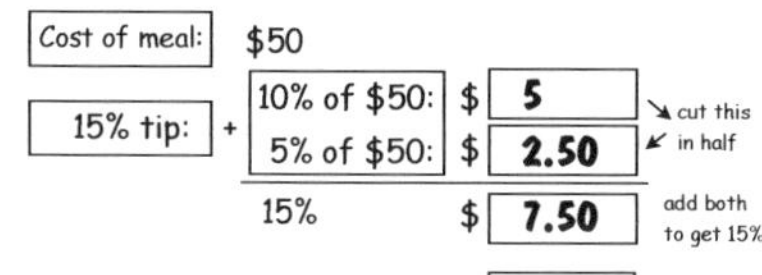

Cost of meal: $50

15% tip: + 10% of $50: $ **5** (cut this in half)
5% of $50: $ **2.50**
15% $ **7.50** (add both to get 15%)

Cost of meal plus the tip: **$57.50**

15% Restaurant Tip Practice

Calculate the cost of the meal plus a 15% tip

1. Calculate the cost of a $140 meal with a 15% tip

Cost of meal: $140

15% tip: + 10% of $140: $ **14** (cut this in half)
5% of $140: $ **7**
15% $ **21** (add both to get 15%)

Cost of meal plus the 15% tip: $ **161**

2. Calculate the cost of a $100 meal with a 15% tip

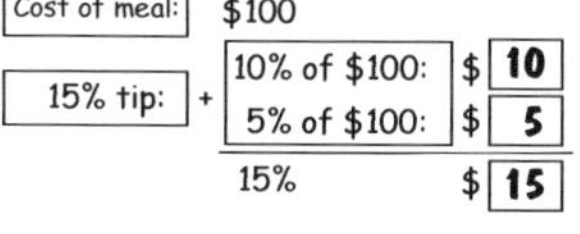

Cost of meal: $100

15% tip: + 10% of $100: $ **10**
5% of $100: $ **5**
15% $ **15**

Cost of meal plus the 15% tip: $ **115**

3. Calculate the cost of a $30 meal with a 15% tip

Cost of meal: $30

15% tip: + 10% of $30: $ **3**
5% of $30: $ **1.50**
15% $ **4.50**

Cost of meal plus the 15%tip: $ **34.50**

4. Calculate the cost of a $12 meal with a 15% tip

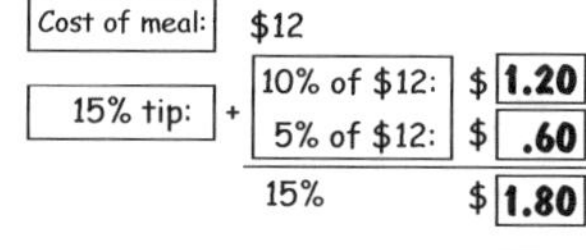

Cost of meal: $12

15% tip: + 10% of $12: $ **1.20**
5% of $12: $ **.60**
15% $ **1.80**

Cost of meal plus the 15% tip: $ **13.80**

5. Calculate the cost of a $180 meal with a 15% tip

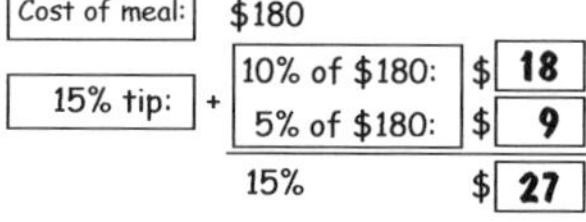

Cost of meal: $180

15% tip: + 10% of $180: $ **18**
5% of $180: $ **9**
15% $ **27**

Cost of meal plus the 15% tip: $ **193.80**

6. Calculate the cost of a $24 meal with a 15% tip

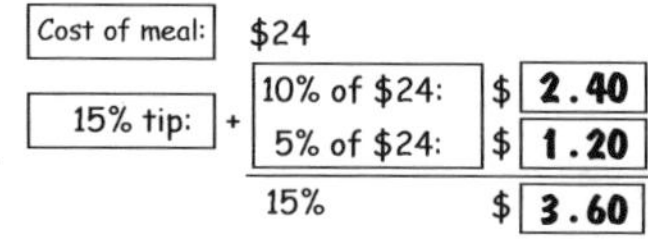

Cost of meal: $24

15% tip: + 10% of $24: $ **2.40**
5% of $24: $ **1.20**
15% $ **3.60**

Cost of meal plus the 15% tip: $ **27.60**

Price With 10% Discount

Example

A. Calculate the cost of a $30 book with a 10% discount

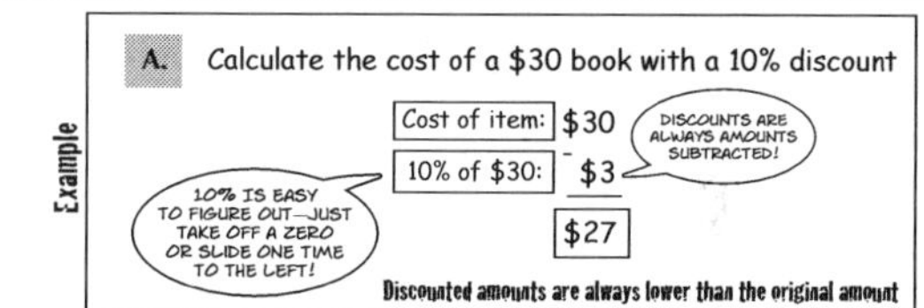

Cost of item: $30
10% of $30: - $3
$27

DISCOUNTS ARE ALWAYS AMOUNTS SUBTRACTED!

10% IS EASY TO FIGURE OUT—JUST TAKE OFF A ZERO OR SLIDE ONE TIME TO THE LEFT!

Discounted amounts are always lower than the original amount

Calculate the cost of the following items with a 10% discount

1. Calculate the cost of a $20 book with a 10% discount

Cost of item: **20**
10%: - **2**
$ **18**

2. Calculate the cost of a $40 pair of shoes with a 10% discount

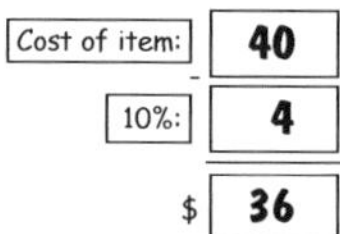

Cost of item: **40**
10%: - **4**
$ **36**

3. Calculate the cost of a $60 jacket with a 10% discount

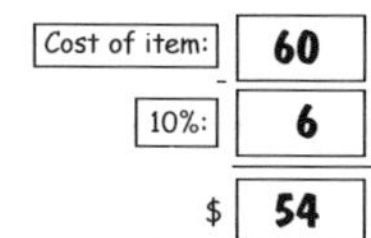

Cost of item: **60**
10%: - **6**
$ **54**

4. Calculate the cost of a $35 sweater with a 10% discount

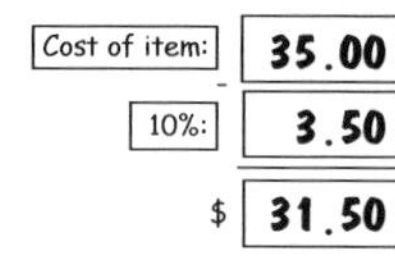

Cost of item: **35.00**
10%: - **3.50**
$ **31.50**

5. Calculate the cost of a $48 gift with a 10% discount

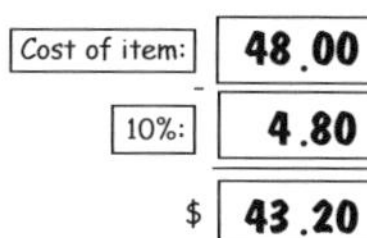

Cost of item: **48.00**
10%: - **4.80**
$ **43.20**

6. Calculate the cost of a $150 gift with a 10% discount

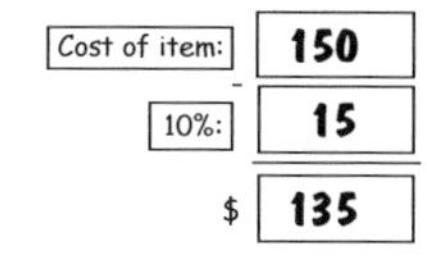

Cost of item: **150**
10%: - **15**
$ **135**

Cost with 10% Tax

Example

A. Calculate the cost of a $20 book with 10% tax

Cost of item: $20
10% of $20: + $2
$22

TAXES ARE ALWAYS ADDED!

Taxes are always added to the original amount

Calculate the cost of the following items with a 10% tax

1. Calculate the cost of a $80 jacket with a 10% tax

Cost of item: **$80**
10%: + **$8**
Cost with tax: **$88**

2. Calculate the cost of a $30 pair of shoes with a 10% tax

Cost of item: **$30**
10%: + **$3**
Cost with tax: **$33**

3. Calculate the cost of a $60 meal with a 10% tax

Cost of item: **$60**
10%: + **$6**
Cost with tax: **$66**

4. Calculate the cost of a $15 shirt with a 10% tax

Cost of item: **$15.00**
10%: + **$1.50**
Cost with tax: **$16.50**

5. Calculate the cost of a $52 pair of shoes with 10% tax

Cost of item: **$52.00**
10%: + **$5.20**
Cost with tax: **$57.20**

6. Calculate the cost of a $12.50 belt with 10% tax

Cost of item: **$12.50**
10%: + **$1.25**
Cost with tax: **$13.75**

Concept Quiz

1. This percent is the same as the whole number ONE: **100%**

2. This percent is the same as the whole number FIVE: **500%**

3. The top number of a fraction automatically tells you **the percent** if the bottom number is **100**.

5. After figuring out a TIP, you take the original amount and then (ADD) / SUBTRACT a percent amount. (circle one)

4. It's common to pay a 15% tip at a restaurant. What is an easy way to calculate 15% of $60? Calculate the tip and the final amount.

Step 1: **Find 10% ($6)**

Step 2: **Find 5% (half of 10%) = $3**

Step 3: **Add the 10% and 5% to find 15% ($9)**

Step 4 (final amount): **Add the 15% to the $60 ($69)**

6. When figuring out a DISCOUNT, you take the original cost and then ADD / (SUBTRACT) a percent amount. (circle one)

7. How would you calculate the price of a $40 item at the store with a 10% discount?

Step 1: **Find 10% ($4)**

Step 2 (final amount): **Subtract this amount from the $40 ($36)**

What is a% of b?

Example

A. What is 32% of 96?

#1 Convert 32% to a decimal (use the DP Trick) — 32% = .32

#2 "of" means TIMES ...so just multiply the decimal times the other number

96 × .32 — Take the decimal off · Multiply normally; · Put the decimal back on at the end!

$$\begin{array}{r} 96 \\ \times\ 32 \\ \hline 192 \\ 2880 \\ \hline 3072 \end{array}$$

3072 → 30.72 (THIS IS THE ANSWER!) Now put on the decimal by sliding two times LEFT

Shortcut! You can also just multiply the percent times the other number and just put two slides on your answer

Calculate the following by converting the percents to decimals and multiplying

1. What is 14% of 82?

#1 Convert the percent to a decimal (use the DP Trick) **14%**

#2 Multiply this new decimal times 82

Answer: **11.48**

2. What is 35% of 75?

#1 Decimal form of the percent: **.35**

#2 Multiply the numbers now:

Answer: **26.25**

3. What is 57% of 63?

Percent as a decimal: **.57**

Answer: **35.91**

4. What is 72% of 45?

Percent as a decimal: **.72**

Answer: **32.4**

Practice with a% of b

ANOTHER WAY TO CALCULATE PERCENTS---
SOME PEOPLE FIND THIS METHOD EASIER

#1 Multiply the percent amount and the number normally

#2 Now give your answer two slides LEFT (because the % has 2 zeros!)

Example

A. What is 42% of 67?

Multiply normally: 42 x 67 = 2,814

Slide your answer to the LEFT 2 times: 28.14

Calculate the following my multiplying normally and giving your answer 2 slides

1. What is 35% of 40?
Multiply normally: **1,400**
Give your answer 2 slides: **14**

2. What is 47% of 90?
Multiply normally: **4,230**
Give your answer 2 slides: **42.3**

3. What is 18% of 75?
Multiply normally: **1,350**
Give your answer 2 slides: **13.5**

4. What is 23% of 68?
Multiply normally: **1,564**
Give your answer 2 slides: **15.64**

5. What is 8% of 36?
Multiply normally: **288**
Give your answer 2 slides: **2.88**

6. What is 3% of 120?
Multiply normally: **360**
Give your answer 2 slides: **3.6**

Practice with a% of b

Calculate the following by converting multiplying and sliding twice to the LEFT

1. What is 20% of 35? **7**
2. What is 63% of 75? **47.25**
3. What is 92% of 80? **73.6**
4. What is 48% of 36? **17.28**
5. What is 18% of 54? **9.72**
6. What is 2% of 87? **1.74**
7. What is 14% of 96? **13.44**
8. What is 3% of 120? **3.6**
9. What is 170% of 38? **64.6**
10. What is 321% of 74? **237.54**

Different Rates of Tax

Example

A. Calculate the cost of a $20 book with 8% tax

$20
× .08
$1.60

CONVERT THE PERCENT TO A DECIMAL AND MULTIPLY!

You can also just multiply by 8 and then give your answer two slides to the LEFT

Cost of item:	$20.00
+ 8% of $20:	$1.60
	$21.60

YOU MAY NEED TO ADD A DECIMAL POINT AND TWO ZEROS!

Use the a% of b method to calculate tax you add

Calculate the cost of the following items with different tax rates

1. Calculate the cost of a $10 binder with 7% tax

Cost of item:	**$10.00**
+ Cost times tax rate:	**$0.70**
	$10.70

2. Calculate the cost of a $30 photo frame plus 8% tax

Cost of item:	**$30.00**
+ Cost times tax rate:	**$2.40**
	$32.40

3. Calculate the cost of a $80 jacket plus 9% tax

Cost of item:	**$80.00**
+ Cost times tax rate:	**$7.20**
	$87.20

4. Calculate the cost of a $12 shirt with 7% tax

Cost of item:	**$12.00**
+ Cost times tax rate:	**$0.84**
	$12.84

5. Calculate the cost of a $15 meal with 8% tax

Cost of item:	**$15.00**
+ Cost times tax rate:	**$1.20**
	$16.20

6. Calculate the cost of a $35 park pass with 6% tax

Cost of item:	**$35.00**
+ Cost times tax rate:	**$2.10**
	$37.10

Finding Totals and Tax

Calculate the cost of the following items with different tax amounts

1. Randy bought two adventure books at $12.50 each and four biographies at $8.50 each. The tax rate was 7%. Calculate the total cost, including tax. How much change would his parents get if they paid with a $100 bill?

$12.50 × **2** =	**$25.00**
+ $8.50 × **4** =	**$34.00**
SUBTOTAL	**$59.00**
+ SUBTOTAL × 7% tax =	**$4.13**
TOTAL	**$63.13**

hundred dollars	**$100.00**
-	**$63.13**
CHANGE	**$36.87**

2. Suzie bought four bags of pet food for $3.10 each and 6 plastic toys for $2.80 each. The tax rate in her city is 6%. Find her total amount.

$3.10 × **4** =	**$12.40**
+ **$2.80** × **6** =	**$16.80**
SUBTOTAL	**$29.20**
+ SUBTOTAL × 6% tax =	**$1.75**
TOTAL	**$30.95**

Round your answer to the nearest 100th

Finding Totals and Tax

Calculate the cost of the following items with a 10% tax

1. Sammy bought two shirts at $4.50 each and three pairs of pants at $10 each. What was the cost of all the clothes and a 10% tax?

SHOW YOUR WORK HERE!

Two shirts at $4.50 each	**$9.00**
+ Three pairs of pants at $10 each	**$30.00**
SUBTOTAL	**$39.00**
+ 10%	**$3.90**
TOTAL	**$42.90**

2. Cindy's mom took the family out to lunch. She bought four hamburgers at $3.25 each and three orders of fries at $1.50 each. How much was the total if there was a 10% tax?

hamburgers	**$13.00**
+ fries	**$4.50**
	$17.50
+ tax amount	**$1.75**
TOTAL	**$19.25**

Percent Review

1. Shade 45%

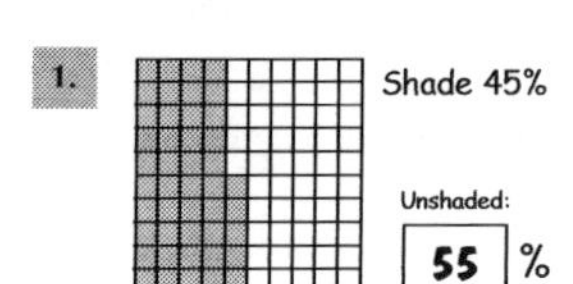

Unshaded: **55** %

2. Divide by 8 = **12.5** %

3. $\frac{12}{25} = \frac{48}{100} =$ **48** %

4. $35\% = \frac{35}{100} = \frac{7}{20}$ (SIMPLIFY!)

5. Calculate the cost of a $70 jacket with a 10% discount **$63**

DECIMAL		PERCENT
6. .02	=	**2** %
7. **1.25**	=	125%
8. .57	=	**57** %
9. **.08**	=	8%

10. 15 out of 50

$\frac{15}{50} = \frac{30}{100} =$ **30** %

11. Calculate the cost of a $50 meal with a 15% tip

$57.50

12. What percent is 15 out of 20?

$\frac{15}{20} = \frac{75}{100} =$ **75** %

13. Calculate the cost of a $12 binder with 6% tax

$12.72

14. Katie bought 3 bags of pet food for $4.00 each and 6 plastic toys for $3.00 each. The tax rate in her city is 6%. Find her total amount.

$42.40

$4 x 3 = $12
$3 x 6 = $18
$30.00
6% x $30 = $1.80
$31.80

Percents Test

1. 20% of 35 =	**7**	**11.** 20% is the same as what fraction? (PUT IN SIMPLEST FORM!)	$\frac{1}{5}$
2. 25% of 24 =	**6**	**12.** 1% of 17 =	**.17**
3. 10% of 70 =	**7**	**13.** 150% of 20 =	**30**
4. 5% of 80 =	**4**	**14.** 70% of 40 =	**28**
5. 50% of 36 =	**18**	**15.** 15% of 80 =	**12**
6. $66\frac{2}{3}\%$ of 12 =	**8**	**16.** 60% = what decimal?	**.60**
7. 200% of 8 =	**16**	**17.** 7% = what decimal?	**.07**
8. 75% of 28 =	**21**	**18.** 25% of 16 =	**4**
9. $33\frac{1}{3}\%$ of 27 =	**9**	**19.** 5% of 40 =	**2**
10. 10% of 43 =	**4.3**	**20.** 10% of .4 =	**.04**

Advanced Percents

- Finding the Percent
- Finding the Whole
- Solving for Percents
- Using the Algebraic Method

Find the Percent

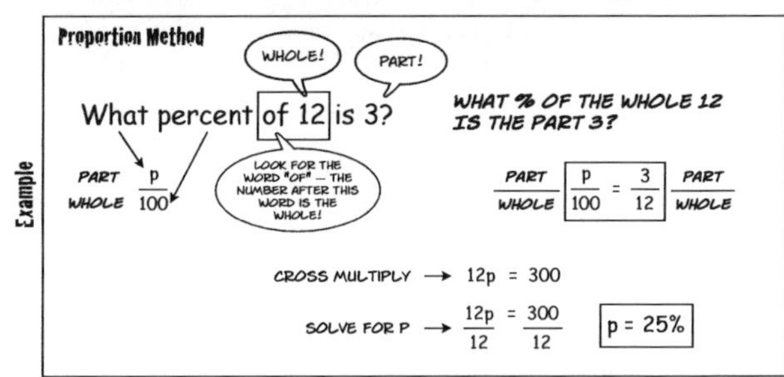

WHAT % OF THE WHOLE 12 IS THE PART 3?

Find the PERCENT, using RATIOS (a.k.a. PROPORTIONS)

1. What percent of 40 is 6? Answer: **15** %

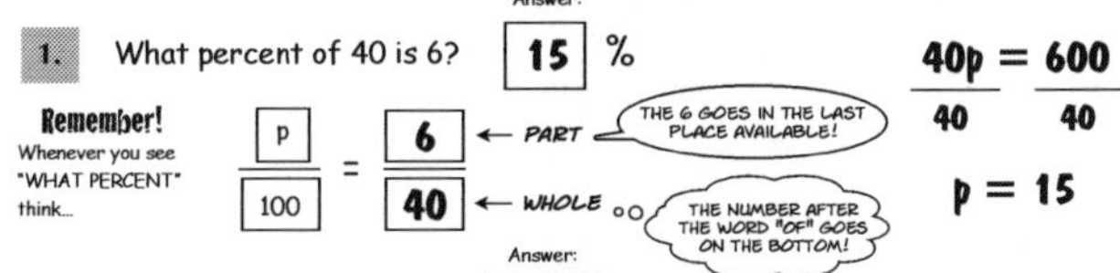

Remember!
Whenever you see "WHAT PERCENT" think...
p = 6 ← PART
100 40 ← WHOLE
THE 6 GOES IN THE LAST PLACE AVAILABLE!
THE NUMBER AFTER THE WORD "OF" GOES ON THE BOTTOM!

$\frac{40p}{40} = \frac{600}{40}$

$p = 15$

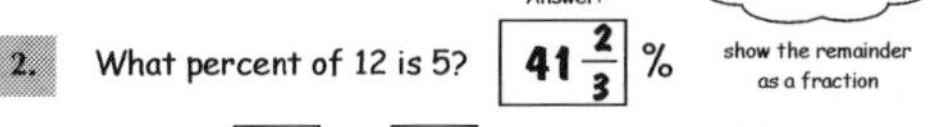

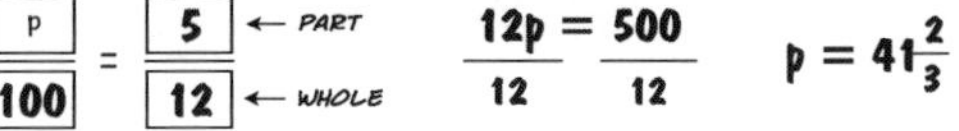

$p = 41\frac{2}{3}$

3. What percent of 5 is 27? Answer: **540** %

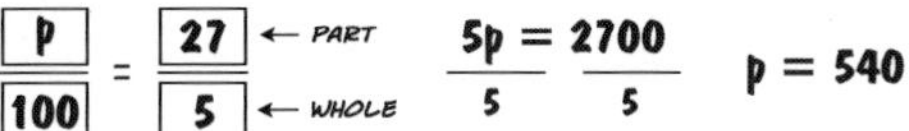

$p = 540$

Find the Percent

Find the PERCENT using PROPORTIONS; round to the nearest tenth

1. What percent of 35 is 7? Answer: **20** %

$$\frac{p}{100} = \frac{7}{35}$$ ← PART / ← WHOLE

2. What percent of 40 is 25? Answer: **62.5** %

$$\frac{p}{100} = \frac{25}{40}$$

3. What percent of 28 is 12? Answer: **42.9** %

$$\frac{p}{100} = \frac{12}{28}$$

4. What percent of 79 is 68? Answer: **86.1** %

$$\frac{p}{100} = \frac{68}{79}$$

5. What percent of 56 is 18? Answer: **32.1** %

$$\frac{p}{100} = \frac{18}{56}$$

Find the % - Fraction, Decimal, Percent Method

Find the PERCENT, going from FRACTION to DECIMAL to PERCENT

Example

A. What percent of 28 is 7?

#1 Make a FRACTION out of the two numbers you have

As before, the number right after the word "of" is the denominator $\frac{7}{28}$

#2 Convert to a DECIMAL by dividing . . .

TOP divided by the BOTTOM $7 \div 28 = .25$

(You may use a calculator on this part)

#3 Convert to a PERCENT by using the DP Trick

Slide the decimal to the RIGHT two times

Stick on the percent sign! You're done! **25%**

	FRACTION (Number after "of" is the denominator)	DECIMAL (Slide this decimal to the RIGHT twice)	PERCENT (Round to nearest tenths place)
1. What percent of 45 is 9?	9/45	.2	20 %
2. What percent of 82 is 56?	56/82	.683 (Round this number to 3 decimal places)	68.3 %

Find the % - Fraction, Decimal, Percent Method

Find the PERCENT, going from FRACTION to DECIMAL to PERCENT

	FRACTION (Number after "of" is the denominator)	DECIMAL (Slide this decimal to the RIGHT twice)	PERCENT (Round to nearest tenths place)
1. What percent of 74 is 23?	23/74	.311 (Round this number to 3 decimal places)	31.1 %
2. What percent of 27 is 94?	94/27	3.481 (Round this number to 3 decimal places)	348.1 %
3. What percent of 65 is 65?	65/65	1	100 %
4. What percent of 40 is 80?	80/40	2	200 %
5. What percent of 32 is 85?	85/32	2.656 (Round this number to 3 decimal places)	265.6 %
6. What percent of 98 is 82?	82/98	.837 (Round this number to 3 decimal places)	83.7 %

Find the Whole

Example

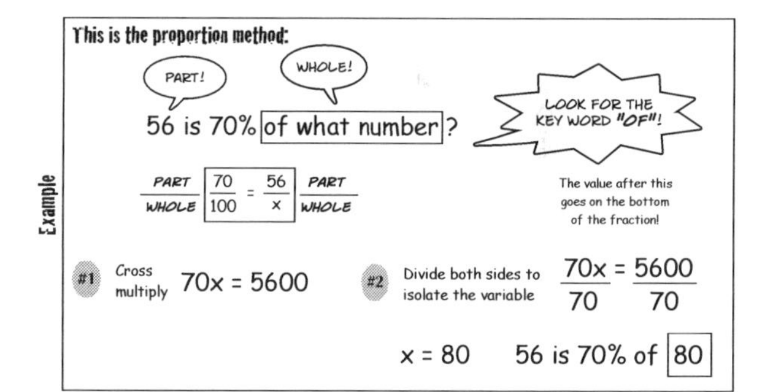

This is the proportion method:
PART!
WHOLE!
56 is 70% of what number?
LOOK FOR THE KEY WORD "OF"!
PART / WHOLE 70/100 = 56/x PART / WHOLE
The value after this goes on the bottom of the fraction!
#1 Cross multiply 70x = 5600
#2 Divide both sides to isolate the variable 70x/70 = 5600/70
x = 80
56 is 70% of 80

1. 32 is 80% of what number? **40**

PART → WHOLE → $\frac{80}{100} = \frac{32}{x}$ $\quad \frac{80x}{80} = \frac{3200}{80} \quad x = 40$

2. 14 is 20% of what number? **70**

$\frac{20}{100} = \frac{14}{x}$ $\quad \frac{20x}{20} = \frac{1400}{20} \quad x = 70$

3. 12 is 8% of what number? **150**

$\frac{8}{100} = \frac{12}{x}$ $\quad \frac{8x}{8} = \frac{1200}{8} \quad x = 150$

Find the Whole

Find the WHOLE, by setting up a proportion and solving for x

Example

A. 12 is 78% of what number?

THE NUMBER AFTER "OF" GOES ON THE BOTTOM— HERE YOU PUT A LETTER!

$\frac{78}{100} = \frac{12}{x}$ ← PART / ← WHOLE

#1 Set up a proportion

#2 Cross multiply 78x = 1200

#3 Get the letter by itself by dividing both sides by the number in front of the letter! 78x = 1200 (÷ 78) (÷ 78) x = 15.38

#4 Round to the nearest tenth Answer: 15.4

1. 37 is 65% of what number? $\frac{65}{100} = \frac{37}{x}$ ← PART / ← WHOLE

#1 Set up a proportion

#2 Cross multiply **65x = 3700**

#3 Get the letter by itself by dividing both sides by the number in front of the letter! (÷65) (÷65) x = **56.9**

#4 Round to the nearest tenth Answer: **56.9**

Find the Whole

Find the WHOLE, by setting up a proportion and solving for x

1. 84 is 145% of what number? $\frac{145}{100} = \frac{84}{x}$ ← PART / ← WHOLE

$\frac{145x}{145} = \frac{8400}{145}$ **x = 57.9**

Answer: **57.9** Round to the nearest tenth

2. 25 is 35% of what number? $\frac{35}{100} = \frac{25}{x}$

$\frac{35x}{35} = \frac{2500}{35}$ **x = 71.4**

Answer: **71.4** Round to the nearest tenth

3. 57 is 142% of what number? $\frac{142}{100} = \frac{57}{x}$

$\frac{142x}{142} = \frac{5700}{142}$ **x = 40.1**

Answer: **40.1** Round to the nearest tenth

Percents Test

1. 20% of 35 = **7**

2. 75% of 32 = **24**

3. $66\frac{2}{3}$ % of 12 = **8**

4. Write 60% as a decimal **.6** or **.60**

5. Write 7% as a decimal **.07**

6. 200% of 9 = **18**

7. What is 3% of 120? **3.6**

8. Multiply by $\frac{1}{3}$ = **33.3** %

9. Calculate the cost of a $40 pair of shoes with a 10% discount **$36**

10. Calculate the cost of a $70 meal with a 15% tip **$80.50**

10% of 70 = 7
5% of 70 = half of 7 = 3.50
15% of 70 = 7 + 3.5 = 10.50
$70 meal + $10.50 tip = $80.50

11. Sammy bought four shirts at $4.50 each and three pairs of pants at $10 each. What was the cost of all the clothes and a 10% tax? **$52.80**

4 x $4.50 = $18
3 x $10 = $30
$18 + $30 = $48 for everything
$48 + $4.80 tax = $52.80

12. Calculate the cost of a $20 binder with 6% tax **21.20**

$20 + $1.20 tax = $21.20

Calculating Percent of Change

Change Over Original

Example

A. Tommy shot 10 baskets at the basketball court on Monday. On Wednesday he shot 15 baskets. What was the percent increase?

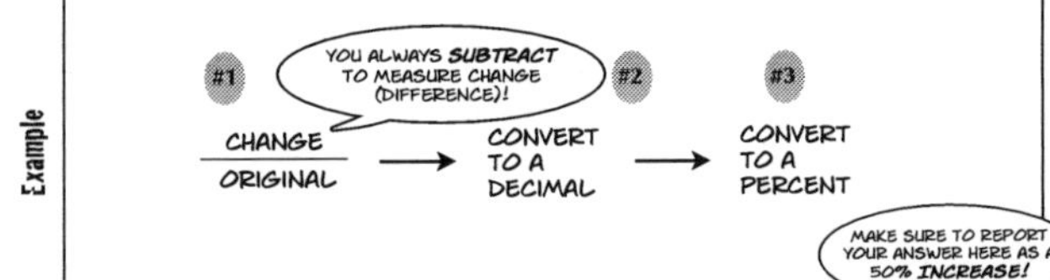

$\frac{15 - 10}{10} = \frac{5}{10} =$ → .5 → 50%

Answer: Wednesday's performance was a 50% increase.

Example

B. Bob used to read 50 pages a night. Now he's reading 40 pages a night. What is the percent of change?

$\frac{50 - 40}{50} = \frac{10}{50} = \frac{1}{5}$ → .2 → 20%

5 ⟌ 1.00

USE THE DECIMAL-HAT TRICK TO CONVERT FRACTIONS TO DECIMALS!

Remember! The top number of a fraction wears the hat!

Answer: Bob now reads 20% LESS than he did before.

IF THE AMOUNT IS DECREASING, BE SURE TO REPORT IT AS SUCH!

IN THIS CASE, USE "DECREASE" OR "LESS"!

Calculating Percent of Change

Change Over Original

1. On Wednesday the temperature was 80 degrees. On Thursday the temperature went up to 100 degrees. What % did it go up?

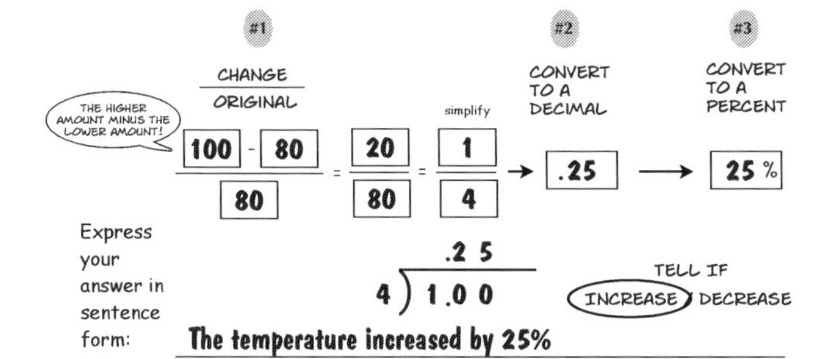

Express your answer in sentence form: **The temperature increased by 25%**

2. Last year Brandon earned $15 an hour at his job. At his new job he makes only $9 an hour. What is the percent of change?

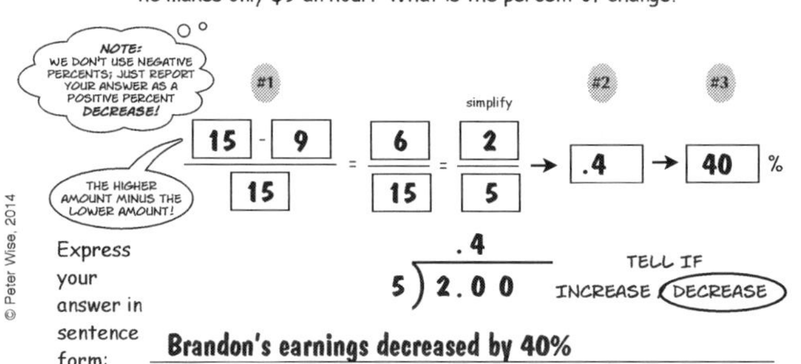

Express your answer in sentence form: **Brandon's earnings decreased by 40%**

Calculating Percent of Change

1. Suzie used to finish her chores in 40 minutes. This week she took 45 minutes to finish the chores. What is the percent of change?

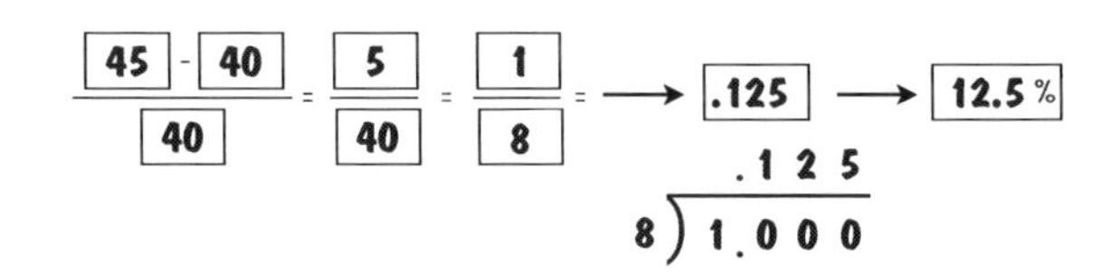

FASTER / SLOWER (circled)

Express your answer in sentence form: **Suzie was 12.5% slower in finishing her chores**

2. Sandra's cleaning business made $160 on her first weekend. The second weekend she made $170. What percent increase or decrease did she have during the second week of operation?

$$\frac{170-160}{160} = \frac{10}{160} = \frac{1}{16} = .0625 = 6.25\%$$

INCREASE (circled) DECREASE

Express your answer in sentence form: **Sandra's business had a 6.25% during the 2nd week**

Turning Fractions into Percents

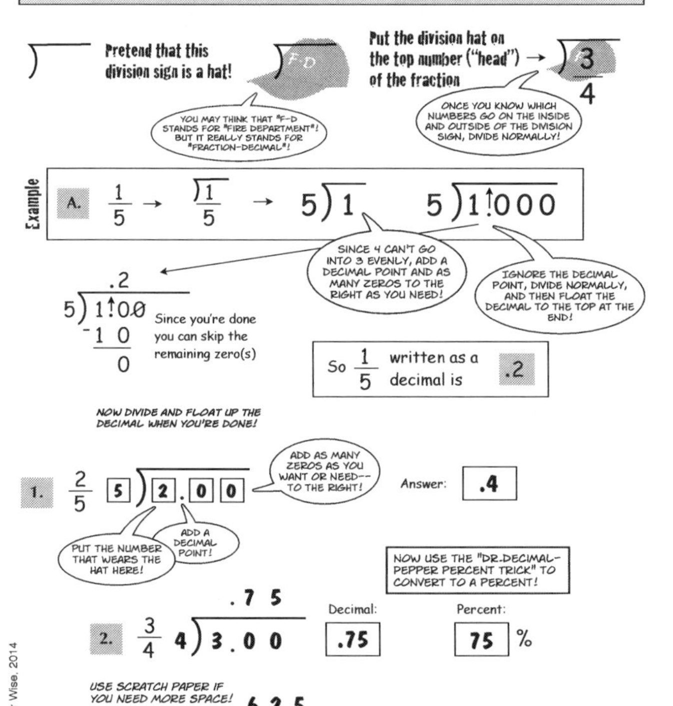

2. $\frac{3}{4}$ 4) 3 . 0 0 = . 7 5   Decimal: **.75**   Percent: **75** %

USE SCRATCH PAPER IF YOU NEED MORE SPACE!

3. $\frac{5}{8}$ 8) 5 . 0 0 0 = . 6 2 5   Decimal: **.625**   Percent: **62.5** %

Mixed Review Percent Problems

Solve the following percent problems; round to the nearest tenths place

CALCULATOR OKAY ON ALL BUT #1 AND #6

1.	87% of 50 =	**43.5**
2.	What percent of 72 is 20?	**27.8**
3.	42% of what is 36?	**85.7**
4.	What percent of 37 is 24?	**64.9**
5.	92 is what percent of 65?	**141.5**
6.	.23% of 358 = **.23% = .0023**	**.8**
7.	73 is what percent of 16?	**456.3**

Solving Percents, using the Algebraic Method

Find the Percent

This is one of three basic types of percent problems. There are two ways of doing these:

- The proportion method (quicker, but easier to make a mistake with these)
- The algebraic method (longer, but some people find these easier to visualize)

Example

This is the algebraic method:

WHOLE! PART!

What percent of 75 is 15?

variable ÷ 100 times 75 equals 15

$\frac{p}{100} \cdot (75) = 15$

#1 Multiply both sides by 100 to get rid of the denominator

$\frac{p}{100} \cdot (75) = 15 \longrightarrow p(75) = 1500$ (• 100)

#2 Divide both sides by the number multiplying the variable

$p(75) = 1500 \longrightarrow \frac{p}{75} = \frac{1500}{75}$

DIVIDE BOTH SIDES BY 75!

THIS IS THE PERCENT!

$p = \frac{1500}{75} = 20 =$ **20%**

Solve the following percent problem

1. What percent of 50 is 4?

$\frac{p}{100} \cdot 50 = 4$ $\quad \frac{p}{2} = 4$ $\quad = 8$ % (Answer:)

#1 CANCEL FROM THE FRACTION AND THE 50!

#2 MULTIPLY BOTH SIDES BY THE NUMBER IN THE DENOMINATOR

Find the Percent

Calculate the following percent problems

1. What percent of 24 is 18? **75** %

$\frac{p}{100} \cdot 24 = 18$

#1 MULTIPLY BOTH SIDES BY 100!

#2 THEN DIVIDE BOTH SIDES BY THE NUMBER MULTIPLYING THE VARIABLE!

$\frac{24p}{24} = \frac{1800}{24}$

$p = 75$

CALCULATOR OKAY ON THESE

Round to the nearest hundredth

2. What percent of 80 is 25? **31.25** %

$\frac{p}{100} \cdot 80 = 25$

MULTIPLY BOTH SIDES BY 100!

THEN DIVIDE BOTH SIDES BY THE NUMBER MULTIPLYING THE VARIABLE!

$\frac{80p}{80} = \frac{2500}{80}$ $\quad p = 31.25$

Round to the nearest hundredth

3. What percent of 90 is 15? **16.67** %

$\frac{p}{100} \cdot 90 = 15$

$\frac{90p}{90} = \frac{1500}{90}$ $\quad p = 16.67$

4. What percent of 40 is 35? **87.5** %

$\frac{p}{100} \cdot 40 = 35$

$\frac{40p}{40} = \frac{3500}{40}$ $\quad p = 87.5$

Find the Percent

Calculate the following percent problems using the ALGEBRAIC method

CALCULATOR OKAY ON THESE PROBLEMS

ROUND THE PERCENTS TO THE NEAREST TENTH!

1. What percent of 40 is 50? **125** %

$\frac{p}{100} \cdot 40 = 50$ $\quad \frac{40p}{100} = 50$ $\quad 40p = 5000$

2. What percent of 48 is 34? **70.8** %

$\frac{p}{100} \cdot 48 = 34$ $\quad \frac{48p}{100} = 34$ $\quad 48p = 3400$

3. What percent of 75 is 20? **26.7** %

$\frac{p}{100} \cdot 75 = 20$ $\quad \frac{75p}{100} = 20$ $\quad 75p = 2000$

4. What percent of 64 is 52? **81.3** %

$\frac{p}{100} \cdot 64 = 52$ $\quad \frac{64p}{100} = 50$ $\quad 64p = 5200$

5. What percent of 84 is 37? **44.0** %

$\frac{p}{100} \cdot 84 = 37$ $\quad 84p = 3700$

Find the Whole

Example

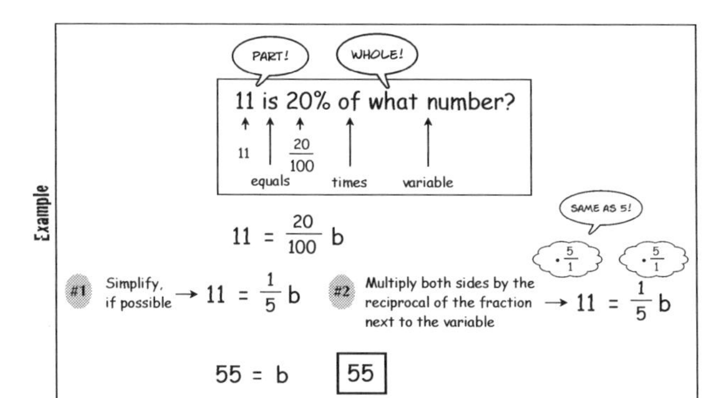

1. 6 is 25% of what number? **24**

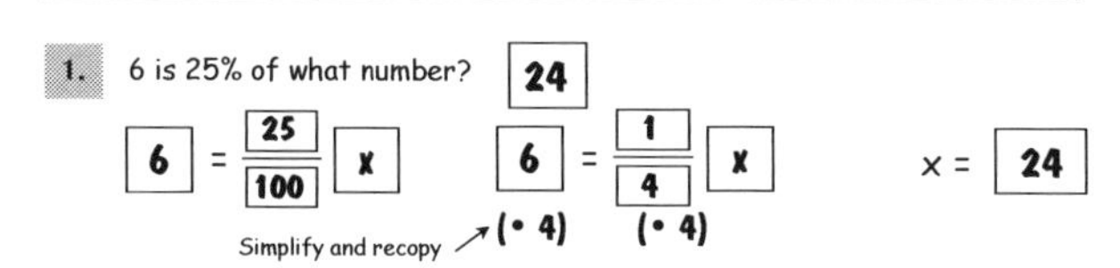

2. 12 is 75% of what number? **16**

$12 = \frac{75}{100}x \qquad 12 = \frac{3}{4}x \qquad x = 16$

Simplify and recopy $(\cdot\frac{4}{3})\quad(\cdot\frac{4}{3})$

3. 9 is 15% of what number? **60** $(\cdot\frac{20}{3})\ (\cdot\frac{20}{3})$

$9 = \frac{15}{100}x$

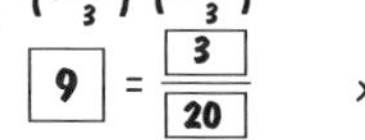

$9 = \frac{3}{20}x \qquad x = 60$

Find the Whole

Example

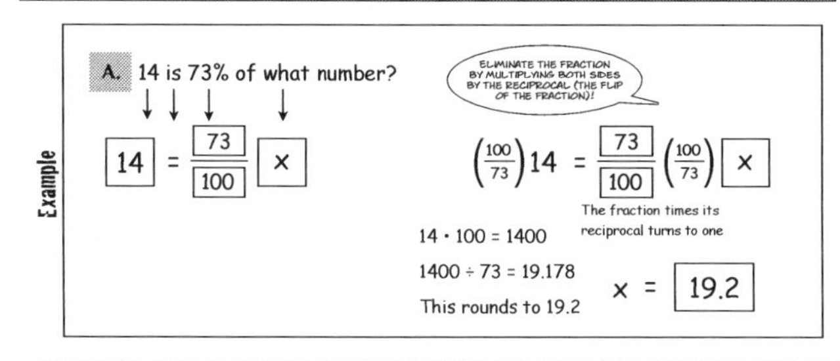

A. 14 is 73% of what number?
ELIMINATE THE FRACTION BY MULTIPLYING BOTH SIDES BY THE RECIPROCAL (THE FLIP OF THE FRACTION)!
14 = 73/100 x
(100/73) 14 = 73/100 (100/73) x
The fraction times its reciprocal turns to one
14 • 100 = 1400
1400 ÷ 73 = 19.178
This rounds to 19.2
x = 19.2

Calculate the following percent problems using the ALGEBRAIC method

1. 8 is 40% of what number?

$8 = \frac{40}{100}x \qquad x = 20$

IT IS A GOOD IDEA TO SIMPLIFY THE PERCENT FRACTION!

ELIMINATE THE FRACTION BY MULTIPLYING BOTH SIDES BY THE RECIPROCAL (THE FLIP OF THE FRACTION)!

2. 27 is 45% of what number?

$27 = \frac{45}{100}x \qquad x = 60$

3. 58 is 110% of what number?

Calculator recommended on this problem
Round to the nearest tenths place

$58 = \frac{110}{100}x \qquad x = 52.7$

Find the Whole

Solve the following percent problems using the ALGEBRAIC method

CALCULATOR OKAY ON THESE

ROUND THE PERCENTS TO THE NEAREST TENTH!

1. 6 is 20% of what number?

$\frac{20}{100} = \frac{1}{5} \qquad 6 = \frac{1}{5}x \qquad x = 30$

Simplify and the percent first

2. 24 is 30% of what number?

$\frac{30}{100} = \frac{3}{10} \qquad 24 = \frac{3}{10}x \qquad x = 80$

Simplify and the percent first

3. 75 is 120% of what number?

$\frac{120}{100} = \frac{6}{5} \qquad 75 = \frac{6}{5}x \qquad x = 62.5$

Simplify and the percent first

4. 72 is 48% of what number?

$\frac{48}{100} = \frac{12}{25} \qquad 72 = \frac{12}{25}x \qquad x = 150$

5. 18 is 14% of what number?

$\frac{14}{100} = \frac{7}{50} \qquad 18 = \frac{7}{50}x \qquad x = 128.6$

Algebraic Method Percents Quiz

Set up equations, using the ALGEBRAIC method, and solve

CALCULATOR OKAY ON THIS PAGE AFTER YOU SET UP THE EQUATIONS

ROUND THE PERCENTS TO THE NEAREST TENTH!

1. What percent of 70 is 20? **28.6** %

$\frac{p}{100} \cdot 70 = 20$

2. 8 is 15% of what number? x = **53.3**

$8 = \frac{3}{20}x$

3. 74 is 168% of what number? x = **44.0**

$74 = \frac{168}{100}x \qquad 74 = \frac{42}{25}x$

4. What percent of 37 is 82? **221.6** %

$\frac{p}{100} \cdot 37 = 82$

5. 26 is 73% of what number? x = **35.6**

$26 = \frac{73}{100}x$

6. What percent of 95 is 71? **74.7** %

$\frac{p}{100} \cdot 95 = 71$

Percents Comprehensive Test

1. 20% of 35 = **7**

2. 80% of 40 = **32**

3. $66.\overline{6}\%$ of 21 = **14**

4. Write the number 11 as a percent: **1100%**

5. Write 48% as a fraction in simplest form: $\frac{12}{25}$

6. Write .6% as a decimal: **.006**

7. Write .095 as a percent: **9.5%**

8. Calculate the cost of a $70 meal with a 15% tip: **$80.50**

9. Calculate the cost of a $35 pair of shoes with a 10% discount: **$31.50**

10. Calculate the cost of a $60 jacket with 5% sales tax: **$63.00**

Percents Comprehensive Test

11. What is 27% of 64? **17.28**

12. What percent of 24 is 5? Express your answer as a mixed number $20\frac{5}{6}\%$

13. 38 is 70% of what number? Express your answer as a mixed number $54\frac{2}{7}$

14. Tony's lawn care business made $150 on his first weekend. The second weekend he made $180. What percent increase or decrease did he have during the second week of operation? **20% increase**

15. What is 2/7 as a percent? **28.6%**
Round to the tenths place

© Peter Wise, 2014

Now that you've learned the basics of percents, look for

MathWise Decimals

or

MathWise Fractions

...with more tips, tricks, and wackiness!

Made in the USA
San Bernardino, CA
22 September 2016